Self-Healing Concrete

David J. Fisher

Published by **Materials Research Forum LLC**
Millersville, PA 17551, USA

Published as part of the book series.
Materials Research Foundations
Volume 101 (2021)
ISSN 2471-8890 (Print)
ISSN 2471-8904 (Online)

Print ISBN 978-1-64490-136-6
ePDF ISBN 978-1-64490-137-3

Distributed worldwide by

Materials Research Forum LLC
105 Springdale Lane
Millersville, PA 17551
USA
http://www.mrforum.com

Printed in the United States of America
10 9 8 7 6 5 4 3 2 1

Table of Contents

Introduction

Concrete is one of the most important and widely-used materials and underpins, literally and figuratively, the entire built environment, thanks to its unique properties. It is however highly susceptible to cracking, due to its inherently low tensile strength, and cracks can result in a marked reduction in operational life plus an increase in maintenance and repair costs. Cracking-related deterioration is one of the most obvious threats to the integrity and durability of concrete structures. The appearance of cracks in reinforced concrete structures, for example, assists the penetration of aggressive chloride ions into the concrete and impairs durability due to the hastened onset of chloride-induced corrosion of the metal reinforcement. Such ions penetrate the concrete matrix along the crack path, and then travel in directions perpendicular to the crack. Repair of the concrete is thus immediately required in order to halt corrosion of the reinforcement so that the crack does not extend further. Incautious repair can lead to unexpected thermal expansion, and also health hazards arising from the chemicals used in the repair. Some 7% of the anthropogenic CO_2 in the atmosphere is due to cement production, and so methods which can prolong the service life of existing concrete structures will also benefit the environment. Due to an increase in the recognition of the negative impact which construction processes have upon the environment, there is now an ever-increasing desire for concrete structures to operate longer while maintaining a high performance level. Cracked concrete must nevertheless be repaired promptly in order to prevent structural damage and prolong structural life. The repair and rehabilitation of concrete structures is expensive, and it may even be difficult to access the damage site following completion of the structure.

Concrete is susceptible to cracking caused by various processes: freeze-thaw cycling, reinforcement corrosion, creep and fatigue. Above all, concrete is a man-made material which comprises cement, coarse aggregates, fine aggregates and water; opening up the possibility of drying-shrinkage. Small resultant cracks have no effect, but large cracks can cause the disintegration of concrete structures. Concrete compositions are chosen so that the development and spread of damage is hindered as much as possible.

The use of continuous inspection and maintenance regimes unfortunately attracts high material and labour costs. Maintenance of the reliability of the concrete infrastructure is strategically important, but the huge associated costs of inspection, maintenance, repair and ultimate replacement are no longer sustainable. One underlying problem is that the design of the concrete infrastructure remains too traditional, and poor material properties are the main cause of the failure or deterioration of the infrastructure.

The repair of concrete structures worldwide, which have been damaged by water or aggressive water-based solutions, is estimated to cost billions of dollars each year, but the treatments which can render concrete structures more durable are limited in number. The use of crystalline admixtures offers the possibility of improving durability, and reducing the permeability of concrete structures exposed to corrosive environments is one possible route.

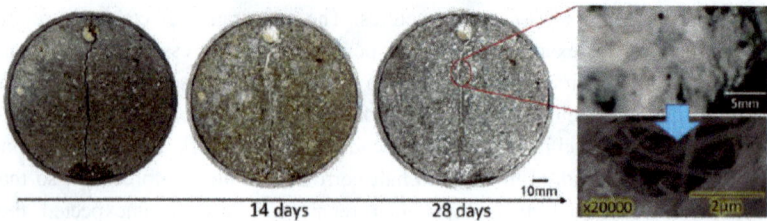

Figure 1. Autogenous self-healing of mortar after 14 and 28 days of curing. Reproduced from Developing the Solution of Microbially Induced CaCO₃ Precipitation Coating for Cement Concrete, Huynh, N.N.T., Nhu, N.Q., Son, N.K., IOP Conference Series - Materials Science and Engineering, 431, 2018, 062006 under Creative Commons Licence 3.0

Autonomous repair phenomena have instead become the most promising path towards side-stepping unsustainable labour-intensive maintenance work. If the harmful cracks could heal themselves, with no human intervention being required, labour costs would evaporate. Self-healing concrete has consequently attracted a great deal of attention during the past two decades. Self-healing not only protects the concrete matrix, but also any steel reinforcement. As a rule-of-thumb, cracks with sizes of up to 0.1mm heal autogenously while cracks with sizes of up to 1mm heal autonomously. Concrete which contains self-healing materials features in many sustainable structures because of the associated decreased maintenance costs and extended service life, although such concrete formulations may require extensive water-exposure in order to guarantee the promised crack-healing. Autogenous healing is produced mainly by continuing hydration or carbonation. As a typical example, the crack-closing produced by autogenous healing (figure 1) of early-age concrete has been evaluated for crack sizes of 0.1 or 0.4mm under conditions of water immersion, exposure in a humidity chamber and subjection to wet/dry cycling[1]. The crack closing was evaluated after 7, 14, 28 and 42 days and the internal status of the cracks was checked visually and by using phenolphthalein. This showed that specimens which were stored in a humidity chamber did not exhibit healing, whereas specimens which were subjected to wet/dry cycling or water immersion led to the complete closure of cracks which were less than 0.15mm in width. The autogenous healing occurred at a higher rate during wet/dry cycling but involved a higher final efficiency under water immersion conditions. Inspection of the specimen interiors showed that the self-closing occurred mainly at the surface, and carbonation of the crack faces was noted only very near to the specimen surface.

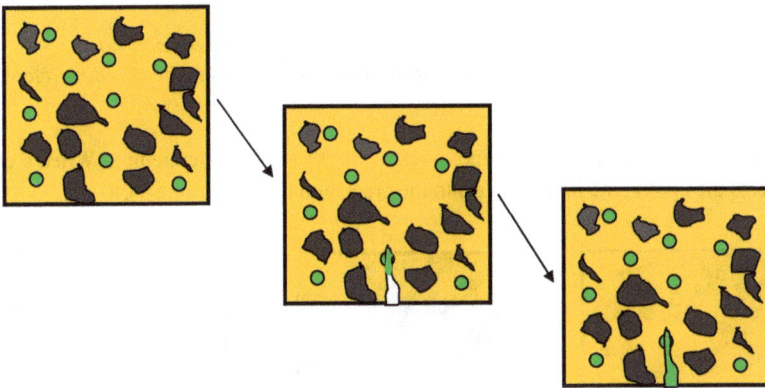

Figure 2. Healing via microcapsule breakage
grey: aggregate, green: adhesive-filled capsule

One autonomous healing method is to embed a repair agent in the concrete during casting, with the agent being loaded into small capsules that break when cracks form (figure 2). The agent is thereby released into the crack, rapidly solidifies and seals the crack. One potential drawback here is that the capsules themselves might impair the strength of the concrete. In a further refinement, 2 components of the healing agent may be dispersed in separate capsules (figure 3).

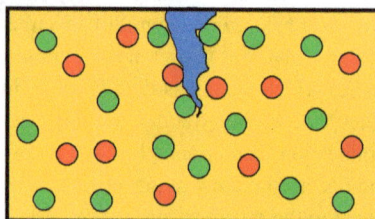

Figure 3. Healing using a two-part microcapsule-based approach
green: monomer plus accelerator, red: initiator, blue: polymerized healing agent

Another strategy is simply to add epoxy resin directly to the concrete mix. When a crack intersects the epoxy, the latter flows into the crack and heals it (figure 4).

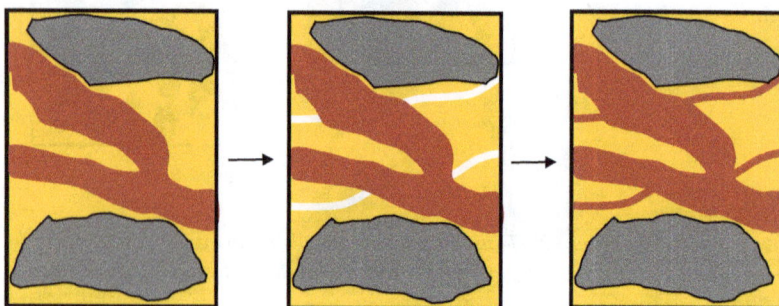

Figure 4. Self-repair by addition of epoxy resin. When a crack intersects the resin reservoir, the low-viscosity resin enters the crack by capillary action and fills it.
grey: aggregate, red: resin

Some materials can heal cracks in concrete by precipitating calcium carbonate. Biological self-healing concrete in particular involves the biochemical reaction of microbe-induced calcium carbonate precipitation. It is essentially a construction material which has been seeded with bacteria and a precursor compound which react and seal cracks when they appear. The bacteria act as catalysts, transforming the precursor compound into limestone. Self-healing concrete containing bacteria which can form $CaCO_3$ crystals for crack-sealing, offers obvious benefits in the form of reduced maintenance costs, with cracks being autonomously repaired without human intervention. Various groups of micro-organisms are able to induce the formation of calcium carbonate, including oxygenic and anoxygenic phototrophic micro-organisms, aerobic organotrophic bacteria (causing ammonification) and some anaerobic micro-organisms (causing sulfate reduction, methane production, denitrification). Little information is available on the use of anaerobic micro-organisms for this purpose.

The drawback here is that concrete is a harsh environment, and the bacteria themselves may require protection. The main issue affecting the self-healing of concrete cracks by microbe-induced mineralization is therefore protection of the bacteria, and this involves the choice of a suitable carrier. In order to prolong the survival of bacteria and improve the self-healing of late-age cracks, core-shell capsule-based healing agents have been used to load spores. In one example[2], self-healing concrete was prepared by mixing fine aggregates with an equal amount of capsule-based healing agents. The capsules could provide protection for loaded spores for at least 203 days. The concentration of calcium ions in the crack-zone solution was also markedly increased. The compressive strength of the concrete did not change greatly for low contents of capsule-based healing agents, while the fluidity of the fresh concrete was much improved. When compared with the strategy of directly adding spore powder, the self-healing ability was clearly improved by adding capsules. Another study[3] used carbide slag, fly-ash and desulfurized gypsum to prepare a cementitious material for coating bacterial spores so as to create biocapsules. The addition of biocapsules as 5% of the cement mass led to the complete healing of cracks with a width of 150 to 550µm. Following healing the permeability of specimens which contained the biocapsules decreased by some 2 orders of magnitude, as compared with control specimens. The use of expanded clay as a carrier imparts bond strength to cement composites as well as protecting the bacteria. A study[4] of the self-healing ability of expanded clay involved styrene-acrylic emulsion coatings and *Lysinibacillus boronitolerans*. Although the coating had a negative effect upon the bacterial density, the latter was higher for coated expanded clay (5.0×10^4cfu [colony forming units]/g of clay) than for uncoated expanded clay (2.4×10^3cfu/g of expanded clay) when exposed to a pH = 12 environment at 60C for 48h. Even with the bacteria within the clay, the bacterial survival rate decreased quickly with time within the mortar. The bacterial density was nevertheless much higher for coated than for uncoated expanded clay after 28 days. The concrete healing rates were 70% for uncoated expanded clay and 75% for coated expanded clay, as compared with rates of 50 and 42% for plain mortar and mortar with empty expanded clay, respectively. Recycled aggregate was used[5,6] as a protective carrier for *Bacillus pasteurii,* and compared with other incorporation techniques such as the direct introduction of bacteria, the use of diatomaceous earth-immobilized bacteria and expanded perlite-immobilized bacteria. The healed crack-widths (0.28mm) of specimens made with recycled aggregate-immobilized bacteria were similar to those (0.32mm) of specimens with expanded perlite-immobilized bacteria, and were larger than those (0.14mm) of specimens with diatomaceous earth-immobilized bacteria or directly introduced bacteria. The interfacial transition zone around aggregates in concrete tends to be full of calcium hydroxide crystals which can act as a calcium source for biomineralization. The possibility of exploiting such zones was explored[7] by using *Sporosarcina pasteurii.* The bacteria were first sporulated, and then protected by fixation in porous lightweight aggregate. The results showed that the use of a lightweight aggregate carrier and the implantation of *Sporosarcina pasteurii* could induce biomineralization, strengthen the interfacial transition zone and repair small internal cracks.

The alkali-carbonate reaction in dolomite aggregate concrete has been considered[8] as a potential self-healing process. Samples were prepared by using a mixture of Portland cement and crushed dolomite aggregate, and were subjected to accelerated testing in 1M NaOH at 60C or in de-ionized water at 60C. This showed that the complete alkali-carbonate reaction was successfully activated, including de-dolomitization and the formation of secondary phases such as $CaCO_3$ and some minor phases which contained Mg-Al, Mg-Si and/or Mg-Al-Si combinations. A so-called carbonate halo and Ca-Al-containing phases also precipitated within deliberately created pre-formed cracks. Cracks which were up to about 200μm in width were completely filled by the carbonate halo and the Ca-Al- phases. Wider cracks were not always completely filled by the newly precipitated phases but their length and width were successfully reduced. Upon filling of the pores and cracks, the mechanical strength of the composite was greatly increased by the alkali-carbonate reaction.

Semi-flexible pavement materials which possess the characteristics of good high-temperature stability, marked durability and low cost are suitable for constructing heavy-duty roads, but the cracking problem has hindered the use of this kind of pavement which would obviously have advantages for the environment in view of their ubiquity. Engineered cementitious composites, for example, have been used[9] to form semi-flexible pavement materials. Test results showed that the fluidity and strength of such cementitious composites met specification requirements when the water/cement ratio was 0.23 and the fiber dosage was 1 to 2%. The flexural strength of the composites was better than that of ordinary mortar. The higher the fiber dosage, the greater was the flexural strength. Increasing the void content of a matrix asphalt mixture and the amount of mortar increased the toughness of the overall material. Curing was the key factor which governed the self-healing properties of the materials. The self-healing effect of materials in 60C water was best.

Cementitious structures tend to be rather crude and bulky and of course susceptible to cracking. In order to counter this, a Voronoi-like tessellation has been used[10] to manufacture small cementitious plates. By using strain-hardening cementitious materials, it was possible to produce thin high-strength forms that could be easily cast. The Voronoi structure imparted a remarkable healing ability.

In the remainder of this work, these various aspects of the concept of self-healing concrete will be considered in more detail.

Theoretical Studies

Cracks greatly impair the performance of concrete structures, and the microcapsule self-healing method is a good means for repairing cracked concrete. In order to determine the optimum microcapsule parameters for ensuring the best healing effect, a 3-dimensional analytical model was proposed[11] which was based upon the microcapsule self-healing mechanism. The model took full account of the radius and volume fraction of the microcapsules, of the anticipated healing efficiency and of the healing probability. A Monte Carlo method was used to check the analytical probability model. In order to deduce further the behaviour of this type of material, models were developed[12] which were based upon extended mechanical variables that simultaneously controlled both the damage and self-healing processes. An uncoupled healing-model which was based upon physicochemical principles was also developed. It applied to concrete-based materials in which the healing mechanism was activated by the precipitation of calcium carbonate within cracks. Self-healing concrete, as a heterogeneous material consisting of cement, sand and capsules containing a healing agent, was studied[13] by using the computational homogenization method. This method can extract the effective properties of heterogeneous materials by using an averaging technique. The macro- and micro-scales can thereby be bridged for the purposes of multi-scale modelling. A so-called representative volume element is used, which is a microscopic cell that contains all of the microphases of the microstructure. When applied to self-healing concrete, the method revealed the influence of the volume fraction of each constituent. In other work, a self-healing composite material was studied[14] with regard to its effectiveness in limiting crack widths in concrete beams that were subjected to sustained loads. A layered-beam numerical model for the transient thermo-mechanical behaviour of reinforced concrete was coupled to a previous numerical model for the transient thermo-mechanical behaviour of a shape-memory polymer. The combined model was supported by experimental data. The model was used to predict ten-year crack widths in standard reinforced concrete beams, showing that the crack-healing composite could exhibit crack widths which were reduced by up to 65%, as compared with an identical beam without crack-healing.

Intrinsic self-healing in concrete was later analyzed[15] by using a more sophisticated, genetic algorithm plus neural network, scheme. The genetic algorithm was used as a stochastic optimizing tool for the network and guided the latter in approaching a global optimum while avoiding its becoming getting trapped at a local optimum. The model was tuned by using a database that was based upon experimental data concerning cement content, water/cement ratio and the effect of additional cementitious materials, crystalline additives and biological healing materials. The degree of self-healing was judged in terms of the predicted crack widths. The model could well reflect the interacting roles played by biochemical materials, silica-based additives and crystalline components. A damage-plasticity constitutive theory of zero-thickness interfaces was directed[16] towards predicting time-dependent self-healing phenomena, based upon fracture-energy concepts. It accounted for the temporal evolution of concrete porosity, as introduced by the self-healing mechanism. The predictions were compared with experimental results for three-point bend tests of crack-opening up to complete failure; before and after various degrees of exposure. Elastic-damage healing models have been used[17] to make a phenomenological study of self-healing materials, by using a semi-analytical model involving a thick-walled cylinder of self-healing concrete. The elastic-damage healing model was based upon the thermodynamics of irreversible processes in the context of continuum damage mechanics. The model used a spectral decomposition of the stress to reflect the behaviours of concrete under tensile and compressive loadings. The Gibbs potential energy was divided into elastic energy, damage energy and healing energy, and damage and healing surfaces were introduced in order to distinguish damage and healing behaviours from the elastic response. An analytical closed-form solution was verified for a thick-walled self-healing concrete cylinder. It was found that, for given values of the healing parameters, the tangential-stress level in an internal radius of a thick-walled self-healing cylinder was more than tripled over that for a thick-walled non self-healing cylinder.

The fracture behaviour of encapsulation-based self-healing concrete was investigated[18] numerically with regard to the effect of capsule volume ratio and core-shell thickness upon the load-carrying ability and fracture probability of capsules. A packing algorithm was used to randomise the meso-scale structure of self-healing concrete with, for a given number of circular capsules of given diameter and shell-thickness, aggregates were generated which had a prescribed distribution of size and volume fraction. The capsules were assumed to be made from polymethylmethacrylate, and crack nuclei were represented by inserted cohesive features having appropriate tension and shear-softening laws operating on all of the boundaries between differing phases. It was shown that the load-carrying ability of the self-healing concrete decreased with increasing capsule volume fraction, while the capsule core-shell thickness-ratio had no appreciable effect upon strength. It nevertheless had a huge effect upon the incidence of capsule-shell breakage. For a given capsule volume fraction, the number of capsules which fractured increased with core-shell thickness-ratio. A general study[19] of solid capsules which represented 5, 10 or 15% of the cement mass showed that the flowability decreased and the air-content increased as the mixing-rate was increased. The compressive, splitting tensile strength and elastic modulus tended to decrease as the mixing-rate of the self-healing capsules was increased. The flowability increased and the air-content decreased as the particle-size of the self-healing capsules increased, while the compressive, splitting tensile strength and elastic modulus tended to increase with increasing particle size.

A simple damage-healing law was proposed[20] which was based upon a time-dependent healing variable that opposed the damage parameter. The model was applied, at the macro-scale, to an isotropic concrete material which was subjected to a tensile load. Coupled and uncoupled self-healing mechanisms were considered, with new healing parameters being defined for each mechanism. A non-linear healing theory was applied to both the coupled and uncoupled self-healing mechanisms and was compared with classical self-healing theory. It was shown that the damage-healing model was able to simulate both coupled and uncoupled healing mechanisms, although non-linear healing theory underestimated the healing efficiency for both the coupled and uncoupled cases, as compared with classical healing theory. Continuum damage-healing mechanics has been used[21] to model self-healing with the aid of a semi-analytical treatment of a concrete beam. An elastic damage-healing model, involving a spectral decomposition technique, was used to investigate the anisotropic behaviour of concrete under both tension and compression. The analytical solution was confirmed by comparison with the conclusions for a simply-supported beam under a uniformly distributed load. The result for a self-healing concrete beam was also compared with that for an elastic one in order to prove the ability of the proposed analytical method to simulate concrete-beam behaviour. It was concluded that, for a given geometry, a self-healing concrete beam could support a 21% greater weight and the total deflection of the beam up to failure was some 27% greater than that for the elastic solution under the ultimate elastic load for both I-shaped and rectangular cross-sections. The critical effective damage was decreased by 32.4% for a rectangular cross-section and by 24.2% for an I-beam, in the case of self-healing concrete.

Table 1. Compressive strengths of concrete cubes before
freeze/thaw, after freeze-thaw and after self-healing in water

Cement (kg/m³)	Aggregate (kg/m³)	Condition	Compressive Strength (MPa)
432	1799	before	62.4
432	1799	after	49.0
432	1799	healed	51.3
409	1782	before	73.7
409	1782	after	52.6
409	1782	healed	55.9
411	1703	before	59.2
411	1703	after	44.5
411	1703	healed	47.6

The self-healing of concrete means that it can also slow the penetration of chloride ions. A meso-scale model of chloride-ion transport was used[22] to study how the ions diffuse into a concrete which exhibits crack self-healing. The model considered zones which contained elements of aggregate, mortar, interface, crack and damage and was checked by comparison with experimental data. Dynamic crack self-healing was then simulated by using moving-mesh and finite-element methods, with the crack and damage-zone sizes varying as a function of time. Microcrack behaviour in encapsulation-based self-healing concrete under uniaxial tension has most recently been analyzed[23] by using finite element methods, assuming the presence of 3-dimensional circular capsules embedded in the mortar. In order to represent crack nuclei, zero-thickness elements with a prescribed traction-separation law were inserted into the meshes. It was shown that mismatch between the fracture strengths of the capsule, of the mortar matrix and of their mutual interface, had a marked effect upon crack nucleation and propagation. A similar fracture strength of the capsule and mortar, combined with a high interface fracture strength, promoted the initiation of a crack in the mortar which then propagated directly into the capsule. This case was the most favourable one for a capsule-based self-healing concrete because a capsule which contained a healing agent was then likely to fracture and thus fulfil its purpose. An interface which had a lower fracture strength than that of the mortar and capsule would generate a crack at the interface and leave the capsule intact and therefore failing to play its intended role.

Table 2. Flow rate through crack as a function of width and temperature

Width (mm)	Temperature (C)	Flow Rate (l/hm)
0.05	20	0.21
0.05	50	0.38
0.05	80	0.60
0.08	20	1.00
0.08	50	1.61
0.08	80	2.90
0.10	20	2.12
0.10	50	3.62
0.10	80	4.87

Micro-crack damage in engineered cementitious composites exhibits self-repair when subjected to wetting-drying cycles, and studies have qualitatively related their stiffness recovery to the crack width. An analytical model that links the stiffness recovery of a single meso-scale crack in engineered cementitious composites to that of a cementitious composite loaded up to multiple micro-crack damage (macro-scale) was investigated[24]. The model successfully predicted the composite stiffness recovery at any strain, on the basis of known macro-scale crack patterns and meso-scale self-healing behaviours, and captured the fundamental phenomena of self-healing in such composites. The interface cracks between the healing agent and a cementitious material, in a self-healing mortar beam under 3-point bending, were also investigated numerically on the basis of an extended finite element method and cohesive surface techniques[25]. After setting up the features of the original crack by using the finite element technique, a parametric study was made of the effects, upon crack initiation and propagation, of the elastic ratio between the self-healing agent and the cementitious material, of the bonding strength and of the fracture toughness of the interface between the self-healing agent and the cementitious material interface. This showed that crack initiation markedly degraded the stiffness of a cementitious material. A flexible healing agent increased the probability of new-crack initiation and healed-crack propagation. A stiffer healing agent produced a stress concentration around the interface, increasing the fracture probability of the interfacial zone.

The influence of self-healing upon the properties of engineered cementitious composites was investigated[26] at multiple length-scales. There was a clear recovery of tensile and flexural properties due to self-healing. At the meso-scale, self-healing led to recovery of the fiber bridging strength back up to about the same level as that of virgin specimens. At the micro-scale, the self-healing was associated with recovery of the frictional bond strength between fibers and matrix, while chemical-bond recovery did not appear to occur. Recovery of fiber/matrix interfacial friction resulted in recovery of the fiber bridging capacity, and this was the underlying cause of the recovery of the mechanical properties of self-healed engineered cementitious composites.

Ultra-high performance concrete contains a large number of unhydrated particles which are available for self-healing in the presence of water[27]. An hydration model was developed in order to simulate the distribution of solids and pores in a cracked microstructure. The effect of the degree of crack self-healing upon chloride-ion transport process was analysed by using a random walk model. This showed that a smaller crack width and a longer healing time were associated with a greater degree of self-healing. It was deduced that portlandite crystallized in the middle of a crack whereas calcium silicate hydrate gel accumulated only on the crack surface. The ion-transport depths in cracked ultra-high performance concrete were simulated by using a finite difference method. With increasing time, the chloride ions migrated not only vertically in the crack but also migrated in the horizontal direction. A lattice Boltzmann single-component model was proposed[28] for the meso-scale simulation of the self-healing that was caused by such continued hydration in a cement paste matrix. This model not only simulated the healing efficiency but also the change in geometry. The simulation revealed that, even when the filling-efficiency was low, some parts of the crack could nevertheless be completely blocked and could thus lead to lower effective ionic diffusivities.

Self-healing cementitious materials based upon urea-formaldehyde microcapsules were investigated[29] by using electrochemical impedance spectroscopy, combined with a single-crack model which treated the relationship between the impedance parameters and the degree of damage. The model was applied to multi-crack specimens, and the coefficients of the model for a multi-crack specimen were determined. It was found that the repair rate of the dynamic elastic modulus and the damage repair rate were not uniform in value, but followed a similar trend. The technique of electrochemical impedance spectroscopy, and a novel equivalent circuit model, has also been used to study the effect of external pressure, curing-time and environment upon the self-healing behaviour of engineered cementitious composites[30]. The electrochemical impedance spectroscopy data for such composites exhibited changes which depended upon the external pressure, curing time and environment. The resistance which was associated with the ion-transport process gradually increased in step with the self-healing process.

The self-healing due to calcium carbonate precipitation in cement-based materials with mineral additives was studied[31] by using a permeability model, combined with a numerical simulation of calcium carbonate formation that was based upon a modified Poiseuille-flow model. The simulations showed that self-healing could be markedly promoted by an increase in pH and Ca^{2+} concentration. The calculated permeability results were consistent with the measurements made of cracks that appeared in the intermediate or later stages of self-healing. This suggested that the model could predict the self-healing rate to some extent. The percentage of calcium carbonate in the healing products was higher for mortar which incorporated only chemical-expansion additives or for cracks which appeared in the later stages, and these parameters more accurately predicted the self-healing rate.

A very general continuum damage-healing model was used to interpret the phenomena which are involved in agent-based self-healing[32]. The plasticity, damage and healing were described in terms of the accumulated plastic strain, a damage variable and a healing variable, respectively. The energy-dissipation and the corresponding kinetic laws of plasticity, damage and healing, respectively, were deduced on the basis of non-equilibrium thermodynamics and the phase field method. Healing was effected by the diffusion of healing agents, which were released by capsules, or of solute atoms. The process was described by diffusion-reaction equations. The model was applied to the simulation of the healing of concentrated or dispersed damage, including: cutting damage, puncture damage, homogeneous damage under a uniaxial tensile stress and inhomogeneous damage under pure bending. Mechanical loading, accumulated damage and the diffusion of healing agents acted together in governing the behaviour of self-healing materials.

Timeline of Experimental Studies

Early experiments[33] on the self-healing of concrete samples studied the damage, in the form of internal cracking, which was caused by rapid freeze-thaw tests. Six formulations of well-cured concrete, in beam form, were damaged to various degrees and then stored in water for 2 to 3 months. Samples which lost as much as 50% of their original relative dynamic modulus, during the freeze/thaw process, recovered almost completely during the storage in water. The compressive strength suffered reductions of 22 to 29% due to deterioration, but recoveries of only 4 to 5% resulted from self healing (table 1). Freeze-thaw tests which were performed on damaged and self-healed specimens, in a partly sealed condition, showed that deterioration was controlled by the ability to take up water.

That is, the greater the amount of water that leaked through plastic foil during freeze-thaw tests, the greater was the degree of deterioration.

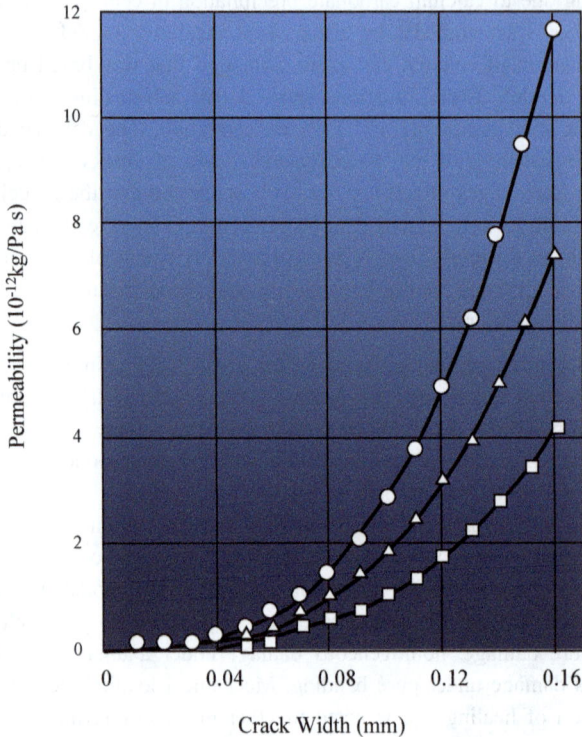

Figure 5. Permeability as a function of crack width
squares: 20C, triangles: 50C, circles: 80C

Later fundamental studies[34] were aimed at developing concretes having a self-healing ability which could lead to strength recovery. Crack-repair agents were here used as so-called 'core' materials in 'shell' bodies that were embedded in concrete structures. The viability of the method was confirmed by using glass pipes which contained a crack-repair agent. Tests were carried out[35] on high-strength concrete in order to determine the permeability and self-healing ability as a function of temperatures of between 20 and 80C, and for crack widths of between 0.05 and 0.20mm (table 2, figure 5). There was an appreciable increase in water transport with temperature. The decrease in the flow-rate depended upon the crack-width and the temperature: smaller cracks healed faster than larger ones, and a higher temperature favoured faster self-healing. With regard to the self-healing of cracks it was concluded that, with an hydraulic gradient of about 1MPa/m and assuming the non-movement of crack edges, 0.10mm cracks could be regarded as smooth and could be closed by self-healing processes.

Table 3. Compressive strength of self-healed high fly-ash content concrete

Fly Ash (%)	Condition	Pre-Load Level (%)	Compressive Strength (MPa)
0	initial	0	51.3
0	initial	70	48.3
0	initial	90	41.3
35	initial	0	46.4
35	initial	70	43.1
35	initial	90	35.6
55	initial	0	38.7
55	initial	70	33.4
55	initial	90	28.1
0	healed, 15 days	0	53.3
0	healed, 15 days	70	53.5
0	healed, 15 days	90	45.9
35	healed, 15 days	0	56.8
35	healed, 15 days	70	51.3
35	healed, 15 days	90	49.8
55	healed, 15 days	0	46.5
55	healed, 15 days	70	43.4
55	healed, 15 days	90	40.2
0	healed, 30 days	0	55.8

0	healed, 30 days	70	56.9
0	healed, 30 days	90	48.3
35	healed, 30 days	0	62.8
35	healed, 30 days	70	57.1
35	healed, 30 days	90	56.3
55	healed, 30 days	0	51.8
55	healed, 30 days	70	50.2
55	healed, 15 days	90	48.1

Due to the continued hydration of cementitious particles, concrete has a natural self-healing tendency following damage, as evidenced by the above-mentioned effect of water immersion. In order to clarify the effect of concrete constituents upon self-healing, factors such as the cement grade and the presence of fly-ash and carbon fibers were studied. This showed[36] that low-grade cement, the replacement of cement by fly-ash and the addition of carbon fibers favoured the self-healing of concrete. The self-healing of concretes damaged at various times before curing was investigated[37], showing that the self-healing of concrete involved crack closure due to new products generated by the continued hydration of unhydrated or insufficiently-hydrated cement particles in the damaged regions. The degree of damage was deduced from the decrease in ultrasonic pulse velocity before and after loading, and the self-healing effect was characterized by using the strength increment following self-healing and introducing[38] a self-healing ratio. There existed a damage threshold in that, when the degree of damage was below the threshold, the self-healing ratio increased with increasing degree of damage. When the extent of damage exceeded the threshold, the self-healing ratio decreased with increasing degree of damage. The damage threshold for normal-strength concrete was higher than that for high-strength concrete.

Table 4. Sorbtivity of self-healed high fly-ash content concrete

Fly Ash (%)	Condition	Pre-Load Level (%)	Sorbtivity (mm/min$^{1/2}$)
0	initial	0	0.113
0	initial	70	0.141
0	initial	90	0.177
35	initial	0	0.084
35	initial	70	0.122
35	initial	90	0.151
55	initial	0	0.089
55	initial	70	0.129
55	initial	90	0.162
0	healed, 15 days	70	0.118
0	healed, 15 days	90	0.139
35	healed, 15 days	70	0.097
35	healed, 15 days	90	0.105
55	healed, 15 days	70	0.099
55	healed, 15 days	90	0.112
0	healed, 30 days	0	0.102
0	healed, 30 days	70	0.108
0	healed, 30 days	90	0.120
35	healed, 30 days	0	0.056
35	healed, 30 days	70	0.067
35	healed, 30 days	90	0.078
55	healed, 30 days	0	0.050
55	healed, 30 days	70	0.055
55	healed, 15 days	90	0.075

One imaginative but apparently short-lived concept[39] was to promote self-healing by incorporating a heating device. The latter, together with a pipe made from thermoplastic organic film containing a repair agent, was embedded in the concrete. Selective heating around a crack could melt the film and allow the repair agent to fill the crack and harden.

The effects of self-healing upon concretes which contained high volumes of fly-ash, when subjected to continuous water exposure, were investigated[40]. Self-consolidating concretes with fly-ash replacement ratios of 0, 35 or 55% were prepared so as to have a constant water/cementitious-material ratio of 0.35. Uniaxial compression was used to generate microcracks in cylindrical specimens of concrete which had been pre-loaded at up to 70 or 90% of the ultimate compressive load. The extent of damage was deduced from the percentage loss in mechanical properties and the percentage increases in the chloride permeability and sorptivity index. Following pre-loading, the specimens were stored in water for a month. The fly-ash mixtures initially exhibited a 27% loss in strength when pre-loaded up to 90% of the ultimate strength. Following 30 days of water-curing, the loss was only 7%, thus indicating the occurrence of appreciable healing. Specimens without fly-ash that were pre-loaded to the same level exhibited an initial 19% loss in strength. Following one month of moist curing, the loss was only 13%. In the case of fly-ash containing samples, the results (tables 3 and 4) were attributed to the self-healing of pre-existing cracks, due mainly to the hydration of anhydrous fly-ash particles on the crack surfaces.

The use of glass tubes having inner and outer diameters of 4 and 6mm, respectively, which were filled with one-part epoxy and embedded in mortar reinforced with steel mesh was essayed[41]. Repeated autonomic healing was possible, with an approximately 30% increase in strength being observed. In order to avoid premature breakage, it was necessary to put a 6.5mm-thick layer of mortar around each tube and cure it for one day before casting.

A fiber-reinforced cementitious composite was developed[42] that required only water and air for self-healing. The main mechanism involved was the formation of calcium carbonate, although the composition also included widely-dispersed minute particles of unhydrated cement that were are available for self-healing. Cement-particle nuclei which were left unhydrated in concrete samples following complete maturation could often impart a self-healing ability to microcracked concrete[43]. The self-healing capability depended upon the amount of unhydrated cement and upon the number and spacing of the particles of unhydrated material. The most important design parameters, with regard to controlling the maximum healable microcrack opening size, were the fineness of the cement and the water/cement ratio. It was noted[44] that self-healing methods had to fulfil at least 6 strict criteria. These included a long shelf-life, quality, reliability, versatility and repeatability. Five types of self-healing method were tested against these criteria: chemical encapsulation, bacterial encapsulation, mineral admixture, chemicals in glass tubing and intrinsic healing based upon a self-controlled limited crack-width. The properties of concrete which had been made from recycled coarse concrete aggregates were such that there was a reduction in the compressive strength, tensile strength, bond strength and porosity, especially at higher recycled aggregate contents[45]. It was suggested that this impairment of properties could be countered by self-healing of the recycled aggregate, by changing the mixing method and by adding silica fume. Self-healing which was produced by immersing the recycled aggregates in water for up to 30 days improved the mechanical properties of recycled-aggregate concrete, especially for low cement contents. The addition of 10% of silica fume, as a cement component, improved the properties.

The effectiveness of crack sealing was evaluated as a function of the presence of crystalline admixtures, maximum crack opening, duration of healing, exposure conditions, fiber orientation and number of cracking and healing cycles[46]. The outcome of self-sealing was expressed in terms of a crack-sealing index. The sealing index, expressed as percentage crack closure, was then correlated with the data from fracture toughness tests which were performed on specimens that were subjected to repeated cracking-healing cycles. The aim was to quantify the maintenance and/or recovery of mechanical properties along the testing path. It was envisaged that a healed crack might re-open and be allowed enough time to re-heal, with this cycle of events being likely to occur several times during the service life. Equivalent tensile stresses, deduced from the absorbed energy per unit of fracture surface, were estimated by using nominal tensile stress versus crack-opening displacement curves of double-edge wedge-splitting tests. Their changes during the cracking and healing cycles were assessed. The results showed that an increase in the sealing index was associated with a slight increase in the toughness as a consequence of through-crack matrix continuity reconstitution and with possibly improved bonding between the fibers and matrix.

Given that concrete can seal its cracks to some extent via autogenous self-healing, a study was made[47] of the autogenous healing of cellulose-reinforced concrete with regard to the self-healing potential and water permeability. Compressive strength and flexural tests were used to measure the mechanical properties and water permeability tests were used to evaluate the coefficient of permeability. Self-healing was monitored by using ultrasonic pulse velocity techniques. The water-permeability coefficient decreased by 42% while the healing-ratio increased at a higher rate in the early days of healing when cellulose fibers were added. There was a 7.8 % increase in the flexural strength and a higher self-healing ratio. A polyvinyl-alcohol fiber-reinforced tailing-sand engineered cementitious composite, has been developed[48] which replaces natural fine aggregate with a large proportion of tailing-sand, thus increasing economy, environmental protection and ductility. In order to explore the self-healing of such composites, cube specimens which contained 50% tailing-sand and 50% natural sand and which had various water/binder ratios were tested with regard to compressive-strength recovery. This showed that, the earlier the composite was damaged, the better was the self-healing response. The self-healing effect increased with increasing self-healing age, but the later growth-rate was lower than the early growth-rate. Wet-dry conditioning was also more conducive to self-healing than were other environments. Self-healing mainly occurred before 21 wet-dry cycles had elapsed.

The effect of mill-rejected granular cement upon the autogenous healing of concrete was explored[49] by using samples which contained 5, 10, 15 or 20wt% of the granular material as a substitute for natural fine aggregate and were aged for 28 days. The specimens were artificially cracked and placed in water or air at 20C in order to cure and heal the cracks for a further 28 days. Compressive, flexural and tensile strength and water permeability tests were used to evaluate the crack-healing. In the case of air-cured specimens, the gain in compressive strength – as compared with control samples - was between 45 and 79%, that in flexural strength was between 74 and 87% and that in tensile strength was between 75 and 84% when the granular material content was between 0 and 20%. In the case of water-cured specimens, the gain in compressive strength was between 54 and 92%, that in flexural strength was between 76 and 94% and that in tensile strength was between 83 and 96% for the same range of compositions. The water permeability coefficients of concrete which was cured in water after cracking decreased by an order of magnitude, while those of concrete which was cured in air increased by the same order of magnitude.

The effects of adding organic microcapsules having various particle sizes and compositions upon the permeability, carbonation resistance, pore structure and self-healing efficiency of cementitious composite material were studied[50] by using rapid chloride migration, water pressure penetration, carbonation and mercury intrusion porosimetry tests. Increasing the microcapsule content to 3% improved the pore structure and impermeability but, at a 6% microcapsule content, the impermeability became lower than that of a specimen without microcapsules. Varying the particle size had a slightly negative effect upon the impermeability but improved the pore structure. Increasing the microcapsule content and particle size improved the impermeability, pore structure and self-healing efficiency.

Electrical measurements can be used to analyze the combined effects of autogenous self-healing and self-sensing in engineered cementitious composites[51]. It has been noted that the electrical methods which have already been extensively used for self-sensing, can also be successfully used to estimate the autogenous self-healing efficiency. The ultimate goal of enhancing the self-sensing ability of such composites for the purpose of accurately monitoring their cracking behaviour and healing[52] was to combine self-sensing and self-healing while maintaining high mechanical strength and ductility. To this end, carbon fibers and carbon nanotubes had been added to cementitious composite mixtures in various concentrations. Compressive and flexural strength, and mid-span beam deflection, tests of non-cracked specimens were used for mechanical characterization. The recovery-rates of flexural properties were assessed using pre-cracked specimens at various healing ages. The combined self-sensing and self-healing abilities of non-cracked and pre-cracked specimens were evaluated using electrical resistivity measurements of specimens having various moisture states and healing ages. Microstructural analysis was used to investigate the outer and inner regions of healed crack-lines. The mechanical properties, self-healing ability and conductivity of the composites could be improved by incorporating carbon nanotubes.

The self-healing of cracked concrete can also be monitored *in situ* by making measurements of diffuse ultrasound. The method was applied[53] to uncracked, tensile through-thickness cracked, and flexural partial-thickness cracked specimens made from 3 different mixes. The test-pieces were unbonded, and post-tensioned to a pre-compression force of 6.2MPa that was chosen so as to generate cracks having widths of less than 200μm and which would exhibit self-healing. The samples were then exposed to a simulated marine environment. In order to monitor the progress of crack healing, measurements were made of the effective ultrasonic diffusivity and of the arrival-time of maximum energy. Measurements were also made of the crack width at the specimen surface. The initial crack provoked an increase in the arrival-time of maximum energy plus an appreciable decrease in diffusivity, as compared with the response of an uncracked control specimen. As self-healing progressed, the arrival-time of maximum energy of the cracked tensile and flexural specimens decreased while the diffusivity increased and returned to the values found for the uncracked control specimens. The changes in the arrival-time of maximum energy and in the diffusivity indicated an autogenous healing of the cracks. It was concluded that the diffusivity predicted the self-healing progress in a more effective way than did the arrival-time of maximum energy.

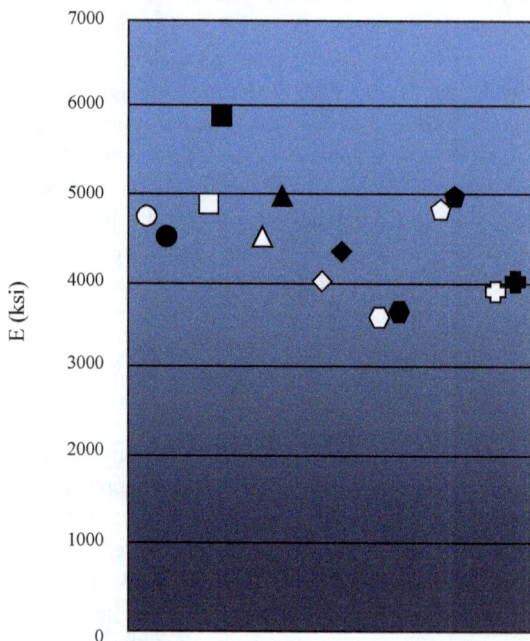

Figure 6. Effect of preparation pH of microcapsules upon the concrete modulus of elasticity before and after healing. White: before healing, black: after healing, circles: control, squares: 0.25%DCPD pH = 3.1, triangles: 0.25%DCPD pH = 3.4, diamonds: 0.25%DCPD pH = 3.7, hexagons: 5%SS pH = 3.0, pentagons: 5%SS pH = 3.1, crosses: 5%SS pH = 3.2

Marine environments affect the self-healing behaviour in various ways. When the coupled effects of external multi-ions and wet-dry cycles in sea-water, upon the evolution of autogenous self-healing in cement paste were investigated[54], it was found that Mg^{2+} rather than Cl^- and SO_4^{2-} in the sea-water played the major role in the healing process. The percentage of brucite in the self-healing reaction products was more than 50%. The closure ratio of a 400μm-wide crack, after healing in sea-water for 56 days, was 2.6 times that for a crack which was healed in tap-water. Wet-dry cycling, using wetting and drying periods of 12h each, could greatly increase the healing efficiency by encouraging the formation of brucite and calcite. The effects of external ions and of wet-dry cycling upon self-healing were more obvious in the case of larger-width cracks because these permitted the ingress of external ions during immersion as well as facilitating the evaporation of water during wet-dry cycles. In related work a self-healing agent was developed which could bind the aggressive ions in marine environments, and simultaneously increase self-healing efficiency, and was encapsulated as artificial aggregate[55]. If the bulk matrix cracked, the artificial aggregates were intersected and the self-healing agent was thus exposed and reacted with synthetic sea-water so as to heal the crack. The reaction of the healing agent with synthetic sea-water, and the formation of Friedel salt (a layered double-hydroxide), ettringite (trigonal calcium aluminium sulphate), and hydrotalcite (another form of layered double hydroxide) in cracks, led to the efficient chemical binding of Cl^-, SO_4^{2-} and Mg^{2+} ions in the synthetic sea-water. For cracks with initial widths of 350 to 450µm, the closure-ratio due to the self-healing was about 45% during the first 12h. This was some 3 times that in control specimens which lacked the healing agent. The overall effect was that aggressive ions which entered cracks could be chemically bound, and the further ingress of aggressive ions from synthetic sea-water was impeded by the rapid crack closure that was due to the self-healing. The possibility of achieving the self-healing of cracked cement paste in sea-water by using triethanolamine was also investigated[56]. Between 0.5 and 2% of triethanolamine (per weight of cement) was added to cement pastes, showing that the triethanolamine could markedly increase the self-healing efficiency in sea-water. In the case of cement pastes which contained more than 1% of triethanolamine, the crack-closure ratio could attain 100% after a healing period of 1 day in sea-water. This was 20 times faster than the rate in cement paste without triethanolamine. The permeability of cracked specimens which contained triethanolamine decreased by some 90% after a healing period of 2 days. In the early stages of triethanolamine-promoted self-healing, large amounts of $Mg(OH)_2$ formed and constituted about 70wt% of the reaction products of the self-healing. Because triethanolamine does not chelate Mg^{2+} ions, the large amounts of $Mg(OH)_2$ which led to the rapid and complete self-healing of cracks the promotion of self-healing by triethanolamine could not be attributed to chelation. In the case of sea-water which contained a high concentration of Mg^{2+}, the ability of triethanolamine to augment the surrounding OH^- concentration led to the formation of $Mg(OH)_2$. This was concluded to be the principal mechanism which led to the promotion on crack self-healing by triethanolamine in cement paste which was exposed to sea-water. The effectiveness of crystalline admixtures as healing agents in chloride-rich environments was investigated[57], with the effect of exposure conditions upon the compressive strength and its recovery in pre-damaged specimens first being analyzed. Crack sealing and the chloride permeability of sealed cracks were evaluated for specimens which were continuously immersed or which were subjected to wet/dry cycling in a 16.5%NaCl aqueous solution. An increased recovery of the compressive strength and an improved crack-sealing ability were observed in the case of samples which contained the healing agent.

An assessment was made of the ability of calcium aluminate phosphate/fly-ash composites to self-heal at an early stage, to re-adhere to surfaces after debonding and to protect carbon steel against brine-caused corrosion before and after re-adherence[58]. Specimens were damaged following autoclaving for 24h at 300C and were then exposed to the same conditions for 5 more days while self-healing and re-adherence occurred. The ability of the material to self-heal and to re-adhere to carbon steel was judged by making measurements of the recovery of the compressive and bond strengths and the Young's modulus, by 3-dimensional micro-image analysis of sealed cracks, by phase identification in the matrix and in cracks after healing and by measuring the carbon steel corrosion prevention by re-adhered composite during a 300C/5-day restoration. Following healing, the sealing of cracks by reaction products and the change in the phase composition of the matrix governed recovery of the original mechanical strength. Analcime production, which arose from pozzolanic autogenous healing, played the main role in sealing cracks, while a well-formed hydroxyapatite phase contributed to the mechanical strength. Re-bonded composite/steel plates exhibited a 0.37MPa bond strength following the 5-day restoration. Electrochemical polarization corrosion tests showed that the re-adhered composite greatly inhibited the brine-induced corrosion of carbon steel and led to a 0.011mm/year corrosion-rate. This was a 50% improvement when compared with the rate before re-adherence. The corrosion inhibition was governed by an improved adhesive contact between the composite and the steel and by increased coverage by the re-adhered composite. An evaluation was made of steel fiber corrosion and the tensile properties of plain and self-healed ultra-high-performance fiber-reinforced concrete when exposed to 3.5% sodium chloride solution[59]. Even after 20 weeks of immersion in NaCl solution, only a few steel fibers which were located close to the surface of non-cracked samples were slightly corroded. These did not noticeably affect the tensile behaviour. A slightly better tensile behaviour resulted from self-healing, and was further improved following exposure to the NaCl solution for longer durations. The surface roughness of pulled-out steel fibers increased due to self-healing and corrosion. This was relevant to the improved tensile behaviour and the effect of increased immersion duration.

A test method was proposed, for the study of the self-healing of cementitious composites[60], which involved evaluating the bonding ability of self-healing products under direct tensile loading. A specially designed cylindrical specimen was split following 7 days of water curing, and the two parts of the split specimen were then immediately put back together. The surfaces of each part were connected to one another with equal pressure for the purpose of healing in water. After 30 days, the specimens were tested using a special direct tension test method in order to evaluate the bond loads. In addition, the crack-closing ratios were monitored on pre-cracked disc-shaped specimens. The bond loads of 7 different ultra-high-performance concrete mixes were compared with the crack-closing ratios. The healing process in the new test, which simulated the self-healing of very narrow cracks, differed from the crack-closing test method. Considerable bond loads were obtained in both fly-ash (308N) and ground granulated blast furnace slag (210N) mixtures. The maximum crack-closing ratio of 100% was observed in the case of the latter mixture.

The self-healing of direct-tension cracks in normal concrete and in engineered cementitious composite panels was investigated[61] under the combined effects of sustained loading and a high water pressure. The panels were subjected to direct tensile forces in order to form full-depth cracks. Water-leakage through the cracks was studied under sustained loading with various water pressures. The leakage rate was recorded over time in order to determine the short- and long-term healing of cracks. The results confirmed that greater self-sealing occurred in the cementitious composite panels, and revealed a marked effect of water pressure upon the self-healing ability of concretes. It was noted that the self-healing process under high water pressures began in the middle portion of cracks and then propagated toward the lower and upper parts.

An ion-chelating agent is a novel crystalline additive which is generally used[62] to improve the self-healing performance of cement-based materials with respect to freeze-thaw damage. The pore-size distribution, structure and compressive strength of mortar which contained an ion-chelating agent were investigated before and after freeze-thaw exposure and after self-healing. The width of a surface crack during self-healing was monitored, showing that the microstructure and mechanical properties of the mortar were improved by the ion-chelation. After 28 days of curing, the proportion of harmful pores (larger than 0.1μm) in mortar which contained 0.5wt% of ion-chelating agent decreased by 42.3%. The compressive strength increased by 26.8%, and many needle-like crystals were observed in the pores; as compared with control samples. Following 100 freeze-thaw cycles, the compressive strength loss, mass loss and harmful-pore content of mortar which contained 0.5wt% of ion-chelating agent decreased by 38.0, 29.2 and 28.7%, respectively, as compared with the equivalent results for control samples. Following curing in water for 28 days, the compressive-strength recovery-ratio of mortar which contained 0.5wt% of ion-chelating agent was 51.8%, with many needle-like crystals being observed in pores and micro-cracks. The main constituents of the needle-like crystals were calcium carbonate, calcium silicate hydrate and ettringite. A surface crack with a width of 0.32mm, in mortar which contained 0.5wt% of ion-chelating agent, self-healed; the self-healing product in the crack being calcium carbonate. Surface cracks having a width of up to 0.4mm in mortar which contained 0.5wt% of chelator, and which were pre-cracked at various ages, could be repaired by standard curing for 30 days. Water permeability tests showed that specimens containing the chelator which were pre-cracked at 3 and 28 days exhibited an excellent self-healing ability with regard to internal cracks, and the relative permeability coefficient of specimens which contained 0.5wt% of the chelator was markedly decreased. Compressive-strength recovery tests showed that the chelator could promote the self-healing of internal cracks. After curing for 56 days, the compressive strength recovery ratio of the mortars with 0.5wt% ion chelator which were pre-loaded at 28 days reached 99.5%[63]. These studies were however limited to fresh-water exposure, leaving it unclear what would happen in sea-water which contained magnesium, sulfate and chloride ions. An investigation[64] was therefore made of the effects of an ion chelator on the pore structure, mechanical properties and self-healing ability of mortar which was exposed to sea-water. The ion-chelator was found to improve the pore structure and compressive strength of mortar under sea-water immersion conditions. Following 90 days of sea-water immersion, the proportion of harmful (greater than 0.1μm) holes in mortar which contained the chelator was 30.3%; lower than that in control samples. The compressive strength following sea-water immersion for 90 days was 34.4MPa, while that of the control was 27.4MPa. The chelator considerably improved the self-healing ability of pre-damaged mortar. Following 28 days of sea-water immersion, the compressive strength recovery-rates of the modified mortar under fresh-water and sea-water conditions were 94 and 80%, respectively. The maximum crack-width which could be completely healed in modified mortar under sea-water was up to 350μm, while the equivalent width for fresh-water conditions was about 240μm. The crack-healing products in the modified mortar under sea-water immersion conditions were magnesium hydroxide and calcium carbonate.

In order to investigate the self-healing characteristics of pre-cracked high-ductility cementitious composites following wet-dry cycles in a marine environment, samples having various water/binder ratios and pre-loading levels were examined[65]. The effect of the self-healing upon the toughness was monitored by using ultrasonic non-destructive and 4-point bending tests. The self-healing products were mainly calcium silicate hydrate gel and $CaCO_3$ crystals, plus some Friedel salt. The recovery-ratio of pre-cracked high-ductility cementitious composites was negatively related to the water/binder ratio and pre-loading level, and positively related to the self-healing age. A pseudo first cracking point kept the cracking process at a relatively high load-level when re-loaded, thus causing the toughness to equal, or exceed, that of control samples after 8 cycles.

A previous effective microcapsule-based self-healing material for marine environments had involved the use of calcium sulfo-aluminate cement as the healing agent[66]. Cracks

Materials Research Forum LLC
https://doi.org/10.21741/9781644901373

were quickly restored by filling with the hydration products which resulted from the interaction of the calcium sulfo-aluminate and water. The degree and rate of self-healing depended sensitively upon the transport of water from the surface to the bulk of the cementitious material. The healing efficiency was up to 82.60%, depending upon the volume of the crack. An investigation[67] was made of the self-healing in various environments of cracks in cement mortar which incorporated metakaolin, bentonite or calcium carbonate microfillers. X-ray computed micro-tomography with 3-dimensional image processing was used to monitor the cracks before and after healing. Following 1 year of exposure, there was no significant self-healing in any of the specimens which were exposed to cyclic temperature and relative humidity changes. On the other hand, all of the specimens which had been submerged in water exhibited various degrees of self-healing, depending upon the type of mineral which had been added. The healing efficiency was 32.26, 27.27, 25.6 and 24.1% for specimens which contained limestone microfiller, Portland cement, bentonite and metakaolin, respectively. Calcium carbonate was the major contributor to the self-healing of surface cracks.

A bio-inspired self-healing method for cementitious materials[68] involved the use of autolytic mineral microspheres in which porous silica with adsorbed sodium silicate was the healing agent and polyvinyl pyrrolidone was the coating material, which could be gradually autolyzed in an alkaline solution. The porous silica microspheres and coating film were prepared by means of sol-gel methods and solvent evaporation, respectively. The autolytic behaviour was found to be controlled mainly by the polyvinyl pyrrolidone thickness, and the dissolution rate of the sodium silicate was affected by the pH value and the temperature. Other polymeric microcapsules containing liquid sodium silicate have been used[69] to produce autonomic self-healing in mortars. This led to improved crack-width and crack-depth reduction and to a recovery of permeability.

In a related context, engineered cementitious composites were considered as vertical cut-off wall materials for the containment of acid mine-drainage[70]. The effect of incorporating reactive MgO into cementitious composites for the enhancement of self-healing was also investigated. The results showed that the hydraulic conductivity of un-cracked and cracked cementitious composites and MgO-bearing composites, pre-strained up to 1.32%, was below the usually accepted limit of 10^{-8}m/s when permeated with acid mine drainage. The self-healing capacity of specimens which were subjected to wet-dry cycles using both tap-water and mine-drainage was improved by the MgO additions. The latter was also beneficial in reducing the hydraulic conductivity of un-cracked and cracked specimens which were permeated with acid mine-drainage. The MgO additions resulted in the formation of new self-healing products, such as hydromagnesite and brucite, when exposed to tap-water, and to the formation of hydrotalcite-like products when exposed to acid mine-drainage. Four-point tests were used[71] to introduce cracks into strain-hardening cement-based composite specimens which contained MgO as an expansive agent. The pre-cracked specimens were then exposed to various curing conditions: dry (50%RH), water-fog (95%RH), tap-water and saturated $Ca(OH)_2$ solution. The addition of MgO could effectively improve the crack-healing ability. Water was essential for the process of crack self-healing, even when an expansive agent was involved, and the healing effect under dry conditions remained poor. When compared with control samples, material which was cured in saturated $Ca(OH)_2$ solution exhibited unsatisfactory crack healing. Strain-hardening cementitious composites exhibit various cracking behaviours when subjected to bending or tensile loads. Specimens were pre-loaded using 3-point bending in order to introduce multiple microcracks and were then exposed to a range of conditions in order to assess their self-healing[72]: water-fog, wet/dry cycling and $Ca(OH)_2$-solution/dry cycling. The self-healing was slow, even in the presence of liquid water or calcium hydroxide solution. The degree of crack-sealing decreased with increasing crack width, with only cracks having widths ranging from 10 to 20μm being completely healed. The $Ca(OH)_2$ solution promoted healing due to enhancement of the pozzolanic reaction of fly-ash. Water capillary absorption could be markedly reduced by crack sealing. The relationship between crack closure and reduced water absorption could be well described by a linear function.

The self-healing performance can be maximized in general by optimizing the temperature and pH so as to control the formation of $CaCO_3$ (figure 7). A study[73] was made of the crystalline form of $CaCO_3$ which was generated during the self-healing of a cement-based composite material. In order to monitor the self-healing as a function of the type of aqueous solution and temperature, the weight-change, weight-change rate and porosity-reduction were determined. In order to increase the generation of $CaCO_3$, nanosized ultrafine CO_2 bubbles were used, together with an adequate supply of Ca^{2+} ions. The latter was achieved by adjusting an aqueous solution of $Ca(OH)_2$. This made it possible to control the production of vaterite having a dense pore structure, together with calcite, by adjusting the temperature and pH value to be about 40C and 12, respectively.

Figure 7. Self-healing precipitates generated at temperatures of 20, 40 and 60C in a constant pH of 9.0. Reproduced from Control of the Polymorphism of Calcium Carbonate Produced by Self-Healing in the Cracked Part of Cementitious Materials, Choi, H., Choi, H., Inoue, M., Sengoku, R., Applied Science, 7, 2017, 546, under Creative Commons Licence 4.0.

The effects of temperature, agitation rate and pH upon the shell thickness and diameter of microcapsules for self-healing were determined[74] for dicyclopentadiene (DCPD) and sodium silicate (SS) healing agents. When the pH was increased from 3.0 to 3.7, the shell thickness increased in the case of sodium silicate whereas the shell thickness attained a minimum at a pH of 3.4 in the case of dicyclopentadiene.

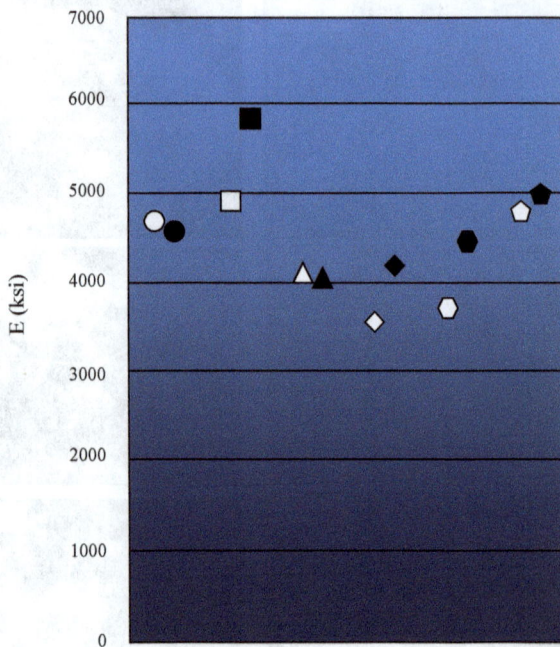

Figure 8. Effect of microcapsules (% of cement weight) upon the modulus of elasticity before and after healing. White: before healing, black: after healing, circles: control, squares: 0.25%DCPD pH = 3.1, triangles: 0.5%SS pH = 3.1, diamonds: 1%SS pH = 3.1, hexagons: 2.5%SS pH = 3.1, pentagons: 5%SS pH = 3.1

The sodium silicate shell thickness was almost twice that of the shell thickness for dicyclopentadiene. The most uniform microcapsules were produced using a temperature of 55C in the case of both agents. For dicyclopentadiene microcapsules, and at temperatures of up to 49C, the solution remained an emulsion and no encapsulation occurred. An increase in agitation-rate decreased the average diameter of the microcapsules in the case of dicyclopentadiene. The diameter of the microcapsules remained constant, for sodium silicate microencapsulation, when the agitation rate was increased from 250 to 550rpm. When sodium silicate was used in self-healing concrete, there was an 11% improvement in the modulus of elasticity when the microcapsules were prepared at a pH of 3.1. The effect of sodium silicate microcapsules upon concrete behaviour was negligible for other pH values. In the case of dicyclopentadiene microcapsules, the increase in the modulus of elasticity of cracked concrete was as much as 30% when the microcapsules were prepared at a pH of 3.1.

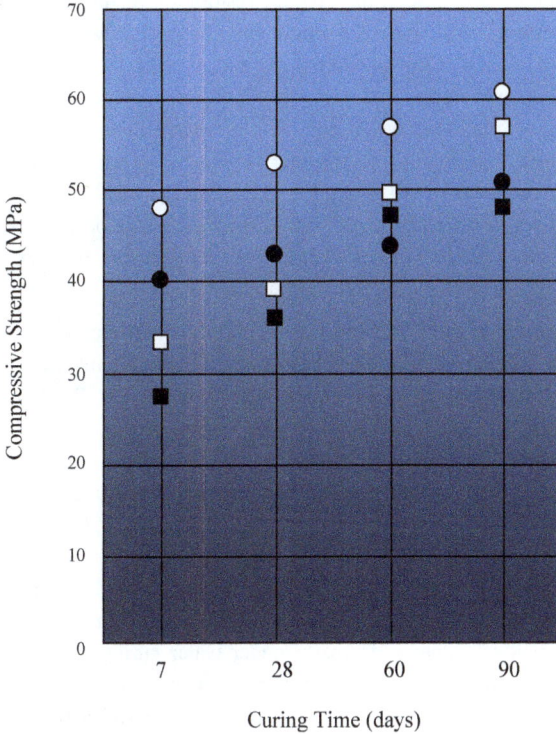

Figure 9. Compressive strength of concrete as a function of curing time and palm-oil fuel ash content. White circles: 0%, black circles: 10%, white squares: 20%, black squares: 30% palm-oil fuel ash

As mentioned before, the presence of water is beneficial to the healing of concrete. A natural healing method involving mineral additives was developed[75] to suit the physical characteristics of an underground environment. The healing ability of a carbonate, a calcium sulpho-aluminate expansion agent and a natural metakaolin was assessed with regard to the relative elastic modulus, strength restoration and water permeability of healed specimens. Specimens which incorporated these mineral additives exhibited differing healing behaviours with regard to improvements in mechanical properties and permeability. The permeability of the host greatly decreased following crack-healing by natural metakaolin, followed by carbonate healing. There was no noticeable improvement in water permeability for specimens which included the expansion agent. Specimens with incorporated carbonate exhibited the greatest restoration of relative elastic modulus and compressive strength. Although the dominant component was $CaCO_3$, arising from reaction with dissolved CO_2 or additives, the healing ability depended upon the morphology and properties of the newly-formed products. A 2-dimensional micromechanical model was developed, for microcapsule-based self-healing concrete subjected to tensile loading in underground structures, in terms of kinetic equations of damage-healing evolution and compliance following healing[76]. A comparison of theoretical and experimental results for a microcapsule-based self-healing mortar confirmed that the model indeed simulated the damage and healing of microcapsule-based self-healing concrete in underground structures.

An investigation was made[77] of the effect of curing conditions upon the self-healing of pre-cracked concrete which comprised palm-oil fuel ash concrete. The palm-oil fuel ash cement replaced 10, 20 or 30% of the total weight of ordinary Portland cement, and the compressive strength was tested after 7, 28, 60 and 90 days of curing in water. Hairline-crack development was monitored under compression. Four environmental parameters were considered when curing: atmosphere, temperature, moisture-content and water presence. Ultra-pulse velocity tests were performed after 7, 28, 60 and 90 days of curing, showing that palm-oil fuel ash concrete, at a 20% replacement level, exhibited the greatest compressive strength. The ultra-pulse velocity increased markedly with increasing curing. The self-healing ability of pre-cracked palm-oil fuel ash concrete, at a 10% replacement level, increased greatly under water curing conditions. The optimum palm-oil fuel ash replacement level which imparted maximum strength was deduced to be 20% (figures 8 and 9), while the inclusion of 10% of palm-oil fuel ash was best for self-healing. The autogenous healing of cementitious materials allows the self-healing of cracks in concrete structures because of the further hydration which occurs when water penetrates the crack and reacts with unreacted binder on the crack face, and because of calcite precipitation which results from the reaction between Ca^{2+} diffusing from the cement paste with the CO_3^{2-} in the penetrating water[78]. Isothermal calorimetry was used to analyze the further hydration of unreacted binder in hardened pastes which contained ordinary Portland cement, supplementary cementitious materials and crystalline admixtures. The amount of heat which was generated decreased, with increasing sample age, because of the reduction in the amount of unreacted binder. Long-term hydration in samples which contained granulated blast-furnace slag and silica fume led to increased heat generation, as compared with ordinary Portland cement. No great difference, in comparison with ordinary Portland cement, was noted for samples which contained fly-ash. When calcium sulfo-aluminate was used as an expansion agent, the heat-generation increased for material which had been aged for 7 days and then decreased after 28 days; again compared with the case of ordinary Portland cement. The products contained calcite, plus an amorphous material, regardless of the binder. Ettringite formed in the specimens which contained calcium sulfo-aluminate and crystalline admixtures. Similar work noted that granulated blast-furnace slag and fly-ash underwent slower reactions with water than did cement[79]. Because of this, there was a high possibility of their being present in an unreacted state in the matrix. Isothermal calorimetry and water-flow tests again indicated that the self-healing potential of ground blast-furnace slag and calcium sulfo-aluminate expansion agents was higher than that of ordinary Portland cement.

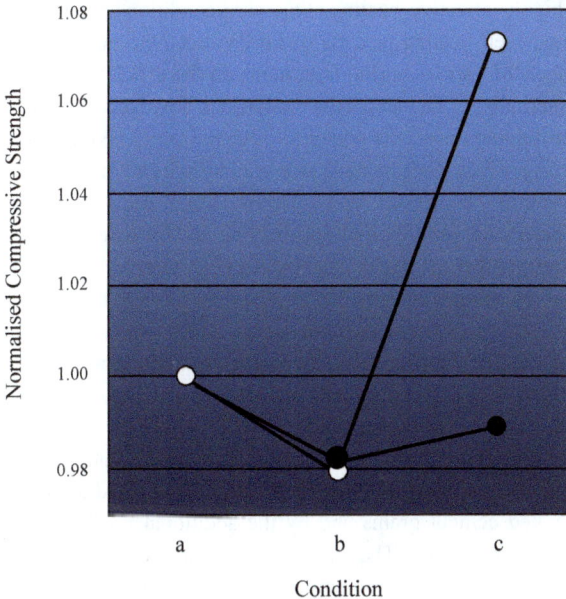

Figure 10. Normalized compressive strength of conventional and self-healing concrete. Black circles: normal concrete, white circles: self-healing concrete, a: initial condition, b: after cracking, c: cracked and submerged in water

It is thus commonly accepted[80] that autogenous self-healing of concrete is controlled mainly by the hydration of Portland cement, with its extent depending upon the availability of anhydrous particles. High-performance and ultra-high performance concretes which contain very high amounts of cement and which have a low water/cement ratio attain degrees of hydration of only 50 to 70%. The presence of a large amount of unhydrated cement should therefore lead to exceptional autogenous self-healing. This hypothesis was checked by performing tests on ultra-high performance concrete and mortar samples having a low water/cement ratio and high cement content. The effects of aging effects were also checked using 1-year old ultra-high performance concrete samples. Analyses were made of the crack surfaces and interiors. The results strongly suggested that the formation of a dense microstructure and rapid hydration of freshly exposed anhydrous cement particles could greatly limit or hinder self-healing. The availability of anhydrous cement therefore did not appear to guarantee the occurrence of highly effective healing. Exposure conditions are thus somewhat critical for the autogenous self-healing process of Portland cement based binders. A study was made of the effects of various curing treatments on the efficiency of crack closure at the surface, and internally[81]. The exposure conditions involved various water-immersion regimes, various temperatures, the presence of chemical admixtures and solutions which contained micro-particles. Self-healing was monitored externally at the surface of the crack, and internally at various crack-depths, using optical and scanning electron microscopy. There was very limited self-healing in the case of pure-water exposure; regardless of the number of cycles, temperature or water volume. The addition of a phosphate-based retarding admixture led to the greatest internal or external crack closure. The highest strength recovery, together with a very good crack-closure ratio, was observed in the case of exposure to water which contained micro-silica particles. The major phase which was detected at the surface was calcium carbonate, while calcium silicate hydrate, calcium carbonate and calcium phosphate were detected internally. Phosphate ions contributed to the filling of the crack, and this was attributed to prevention of the formation of a dense shell composed of hydration phases on exposed areas by unhydrated cement grains and by the additional precipitation of calcium- and phosphate-based compounds. The micro-silica particles were assumed to act as nucleation sites for self-healing product growth. Changes in the chemical composition of the self-healing products occurred as a function of distance from the specimen surface.

A study[82] was made of the optimisation of the mix of expansive minerals and a correlation was established between self-healing and the cracking age of cementitious materials. The degree of hydration of cementitious materials was considered to be a quantitative measure of age. The healing performance was assessed in terms of load recovery, crack-sealing efficiency and gas permeability. The self-healing performance of plain cement mixes increased linearly with a reduction in the degree of hydration. On the other hand, cement mixes which contained an expansive mineral led to a better healing performance than did mixes which contained only Portland cement; regardless of the cracking age. Mixes which contained only Portland cement were associated with healing products that were mainly calcite and portlandite. The optimum use of expansive minerals produced denser healing materials with associated products that included calcium silicate hydrate, and complexes of calcium, magnesium, silicon and aluminium in addition to calcite and portlandite. The proportions of calcite, portlandite and ettringite in the healing products increased with the age of the hardened cementitious material. A further study investigated[83] the effect of expansive minerals such as magnesium oxide, bentonite clay and quicklime on the early-stage autogenous self-healing ability of Portland cement paste. The healing performance was assessed by monitoring flexural strength recovery, crack-sealing and permeability. The hydrated and swelling products of the expansive minerals contributed to the production of healing materials. Cracks with sizes of the order of 180μm were efficiently healed, in mineral-containing mixes, within 28 days. Self-healing was initiated by crack-bridging, leading to strength recovery, and by crack sealing that involved the physical closure of cracks by crystallisation.

A microcapsule-based self-healing system was developed[84] which was aimed mainly at protecting steel reinforcement bars from corrosion, and the performance of the method was monitored via the electrochemical impedance spectroscopy of steel bar that was immersed in simulated concrete. This revealed a marked inhibition of chloride-driven corrosion when microcapsules were added to aqueous solutions containing various mass fractions of sodium chloride. An equivalent-circuit model took account of induction effects arising from the generation of corrosion products on the steel surface, and was used to explain the protective effect of microcapsules against the corrosion of steel bars in concrete. The release of the healing agent from a microcapsule was investigated[85] by means of ethylene diamine tetra-acetic acid titration, showing that the release of a corrosion-inhibitor covered with polystyrene resin was a function of time and was controlled by the thickness of the microcapsule wall. The pH value also affected the release rate of the corrosion inhibitor, with the release-rate markedly increasing with decreasing pH-value. The resistance of polyurethane-containing brittle encapsulation materials to the stresses involved in the mixing and casting of concrete was studied[86], and realistic crack-patterns were created in beams containing embedded capsules. Water-permeability testing of the crack-repair efficiency indicated that there was some improvement in water-tightness, due to self-healing, but the entry of water into the cracks was not completely prevented. The self-healing of dense concrete formulations was investigated by adding coarse cement particles as a self-healing agent[87]. Longitudinal cracks were created in cylindrical specimens, following 28 days of curing, and the specimens were then fastened in such a way that the crack width was fixed at 200μm. The specimens were then cured in water. The secondary tensile strength and the amount of water passing through the cracked specimens were determined, while the degree of hydration was taken to be a measure of the self-healing rate. This showed that the use of coarse cement particles restored up to 38% of the tensile strength, and decreased the rate of water passage by up to 100%. Self-healing mortars containing cementitious tubular capsules housing a polyurethane repair agent were also investigated[88]. Their mechanical performance was evaluated in terms of the recovery of load-bearing capacity under static conditions and of the number of cycles-to-failure as a function of the peak force under cyclic conditions. Positive results were found, with a maximum load-recovery index of up to more than 40% and with the number of cycles-to-failure usually exceeding 10000, while the peak force which was applied during cyclic loading corresponded to at least 70% of the estimated load-bearing capacity of healed samples.

Self-healing in the presence of various concentrations of sodium sulfate solution was investigated[89] with regard to mechanical properties, crack-widths, phase composition and microstructure before and after repair. This showed that self-healing in sodium sulfate solution was much superior to that in water. A 2% sodium sulfate solution imparted the optimum self-healing behaviour at a curing time of 60 days, and the maximum ratio of strength restoration was 126.5%. A 6% solution imparted the optimum self-healing performance in a curing time of 28 days, but the ratio of strength restoration was reduced, from a maximum value of 137.8 to 75.8%, at a curing time of 60 days. The volume of hydration products, such as ettringite, gypsum and calcium silicate hydrate gel, increased with increasing sodium sulfate concentration. Ettringite wrapped in calcium silicate hydrate gel filled the crack and sealed it. When the sodium sulfate concentration was within a certain range, the hydration products could collect and expand, thus generating a second crack. Damage testing of concrete mixed with 30% of fly-ash was carried out[90] using 4 types of pre-loading in the presence of a sulfate environment. The degree of self-healing was judged on the basis of the relative dynamic modulus and compressive strength. The results showed that 5% of sulfate could accelerate recovery of the relative dynamic modulus and compressive strength during the designated curing time. In the case of both water and sulfate environments, 30% of fly-ash could improve the recovery range of the relative dynamic modulus and relative compressive strength only when the degree of pre-load damage was relatively high. With increasing degree of pre-load damage, and in a 5% sulfate environment, the recovery of the relative dynamic modulus and relative compressive strength decreased following mineral self-healing. Following a self-healing curing time of 28 days, the relative dynamic modulus stabilized. The self-healing products in the 5% sulfate environment were mainly calcium carbonate plus some ettringite.

Silica microcapsules which contained an epoxy compound, and silica nanoparticles which were functionalized with an amine group, were studied[91] as a possible self-healing system. The particles were amorphous and were of a suitable morphology and size to be used as cement additives. The amine group exhibited a pozzolanic nature which was better than that of the silica fume which was used as a control. The epoxy compound was quite stable during reaction with lime. It was concluded that these particles were suitable for the production of self-healing concrete.

In order to determine the self-healing and leakage characteristics of a storage tank, the self-healing behaviour of penetrating cracks under various constant-head conditions was studied[92], and leakage-rates were measured with a decreasing water-head. This revealed that self-healing could occur when there was a positive flow of water through the cracks and that a linear relationship existed between the leakage-rate and the decrease in water-head. Studies of a reinforced concrete tank wall under highly eccentric tension revealed a good self-healing behaviour of the cracks[93]. A compression zone which developed in the wall, as a result of flexural stresses, could prevent leakage through the crack. The crack-sealing mechanism is usually attributed mainly to the hydration of anhydrous cement, to the formation and crystallization of calcium carbonate, to swelling of the cement matrix and to the sedimentation of particles within the crack interstices. It has been suggested[94] however that these mechanisms did not explain the sharp water-flow recovery which occurred within the early stages of water permeation through cracked concrete. It was noted that water flow through a narrow crack creates air bubbles via various water-flow mechanisms. These air bubbles then narrow the flow-paths in the crack and lead to an appreciable water-flow reduction.

An *in situ* polymerization method was used[95] to produce double-walled microcapsules which contained sodium silicate as a healing agent within the polyurethane/urea-formaldehyde container. The double-walled microcapsules offered an increased durability at high temperatures, as compared with that of single-walled microcapsules, but maintained adequate interfacial bonding. The microcapsules were incorporated into concrete beams, and microcracks were created by slightly displacing the middle of a beam. A lower pH value and a higher agitation rate and curing temperature improved the formation of microcapsule shells. The healing rate in concrete beams which contained 5% of microcapsules was higher, during the first week, than was the rate in specimens which contained 2.5% of microcapsules.

The use of granules of a self-healing agent made from geo-materials, rather than a self-healing agent in powder form, was considered[96] for the prevention of water leakage through cracks. This approach was based upon the granulation of a self-healing agent, made mainly from geo-materials, which exhibited a high reactivity with water and a long-term retention of self-healing ability. It was found that a 40kg/m^3 admixture of granules, based upon a self-healing agent as a fine aggregate replacement, had a high water-leakage prevention effect due to the self-healing of cracks. The self-healing effect of a crystalline admixture in 4 types of environment was studied[97] by performing permeability tests on cracked specimens. The crack openings were smaller than 300μm, and the time allowed for healing was 42 days. There were differing healing behaviours, depending upon the exposure conditions and upon the presence of the crystalline admixture. The results confirmed that the presence of water was necessary for the occurrence of healing reactions.

The potential use of a calcite-precipitating bacterium as a crack-healing agent was explored[98]. *Bacillus sphaericus* bacteria were prepared in various concentrations and incorporated into concrete mixes. Compressive strength tests were performed after 28 days of curing. Mortar cubes which contained 10, 20, 30, 40 or 50ml of bacteria/cube were prepared. The compressive strength of 150 x 150 x 150mm blocks was good in comparison with that of control samples. When a load was applied to the latter samples, cracking developed earlier. In the case of the bacterial concrete, cracks did not develop in the early stages. A new process[99] for obtaining powder containing an efficient ureolytic microbial agent was developed and this product was as good as the benchmark *Bacillus sphaericus* with regard to urea hydrolysis (20g/l in 24h) and calcium carbonate precipitation. Straightforward incorporation of this material in concrete was efficient at levels of 0.5 to 1% of cement weight. In general, over the entire range of bacterial types, various calcite precipitation mechanisms occur, depending upon the bacterium. The basic differences concern ureolytic ability, with urease-positive bacteria causing rapid and extensive increases in pH. Non-ureolytic strains produce the same changes, but more slowly and more locally. The pH changes are closely related to patterns of precipitation on solid media. Both mechanisms lead to high levels of precipitation, but differences in the precipitate are striking. Ureolytic bacteria produce homogeneous inorganic fine crystals, while the crystals produced by non-ureolytic strains are larger and have a mixed organic/inorganic nature. When tested in regard to crack healing in cement mortars, non-ureolytic bacteria give robust results while ureolytic bacteria lead to more variation. Urease activity leads to the most rapid precipitation, but the same amounts of calcium carbonate are produced by non-ureolytic bacteria and the crystals tend to be larger and have a higher organic content.

Healing agents such as sodium and potassium silicate, acrylic resins and tannins were tested with regard to their adhesive and mechanical properties[100]. Two forms of encapsulation were considered: one involved the encapsulation of sodium silicate solution while the other involved the use of extruded cementitious hollow tubes with various diameters as containers. The latter type of capsule exhibited good mechanical properties and could survive mixing. Three-point bend tests were used to characterize the self-healing performance of the extruded cementitious hollow tubes. Good results were obtained for concrete samples containing tubular capsules filled with sodium silicate, with self-healing being achieved even for cracks that were greater than 1mm. An infiltration crystallisation concentrating agent was prepared[101] by using recycled concrete micropowder as the main component, together with sodium silicate and aluminium stearate. Compression tests were halted when penetrating cracks appeared, and the specimen was then subjected to standard curing for 28 days. The concentrating agent exhibited good self-healing abilities. The optimum mixture for the self-regenerating infiltration crystallisation concentrating agent was 55% cement, 25% quartz sand, 5% sodium silicate, 1% aluminium stearate, 20% recycled concrete micropowder and 8% CaO; with a water/cement ratio of 0.6 and a grey sand ratio of 1/3. Silica fume (0, 2.5, 5, 7.5, 10 or 12.5%) was added to concrete, and the cement was replaced with 35 or 55% of ground granulated blast furnace slag[102]. The specimens were tested in compression after 28 days, and 70% or 90% of the compressive load was applied to another set of specimens in order to generate microcracks for the study of durability. The results indicated that self-healing could be achieved by using silica fume and ground granulated blast furnace slag as admixtures. Those materials also increased the compressive strength. The self-healing of concrete with micro-encapsulated calcium nitrate was investigated after specimens were damaged by applying 80% of their ultimate strength[103]. All of the specimens were incubated by immersion in water. The concentration and size of the microcapsules had a direct effect upon the compressive strength, and specimens into which microcapsules were incorporated had a greater surface resistivity than that of control samples. Although the microcapsules decreased the compressive strength of the concrete, they increased its self-healing ability. More recent work[104] has confirmed that, with *Sporosarcina pasteurii* as the working bacterium, calcium nitrate is a suitable calcium source. Reaction with hydrolysed urea entirely sealed the crack surface and could fill more than half of the crack space. The impermeability of the cracked concrete was largely renewed following repair, and both the frost resistance and sulphate-attack resistance were compensated to some extent.

An attempt was made to arrest cracks in concrete by using bacteria and calcium lactate. The percentages of bacteria used were 3.5 and 5% by weight of cement[105]. Calcium lactate was used, at 5 and 10% by weight, to replace cement. The *Bacillus pasteurii* bacteria produced calcium carbonate crystals which blocked micro-cracks and pores after reacting with calcium lactate. The choice of bacterium depended upon the alkaline environment in which the bacteria had to survive. The bacteria also improved the compressive strength. The bacteria which can produce calcium carbonate include ureolytic bacteria, non-ureolytic bacteria, cyanobacteria, nitrate-reducing bacteria and sulphate-reducing bacteria. The most studied bacterium for this purpose is *Sporosarcina pasteurii*.

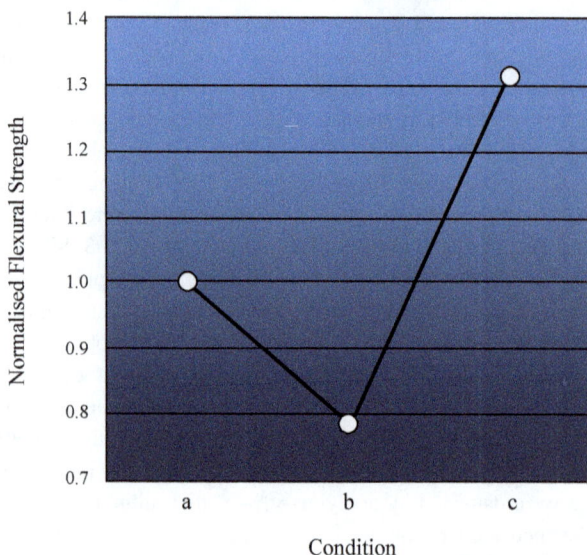

Figure 11. Normalized flexural strength of self-healing concrete
a: initial condition, b: after cracking, c: after healing

A new method[106] was based upon encapsulated polyurethane which was embedded in the concrete, with self-repair occurring when crack formation caused capsule breakage, thus leading to the release and subsequent hardening of the polyurethane within the crack. Another approach involved adding a super-absorbent polymer to the concrete. Such a polymer absorbed any water which entered via a crack, swelled and blocked the crack. When it later released the water, that promoted further hydration and the precipitation of calcium carbonate. Beams (150 x 250 x 3000mm), with and without self-healing

Materials Research Forum LLC
https://doi.org/10.21741/9781644901373

modifications, were subjected to crack creation in 4-point bending. The measured crack-width reduction with time showed that increased autogenous crack-healing occurred when super-absorbent polymers were added. Acoustic emission monitoring confirmed the occurrence of glass-capsule breakage upon crack formation. Other investigations showed that the cracks were partially filled with hydration products, calcium carbonate crystals and/or polyurethane which came from broken capsules. A similar self-healing system was based[107] upon the micro-encapsulation of silica sol via the interfacial polymerization of poly(urea-urethane). Uniform 60-to-120μm spherical capsules were synthesized and embedded. Following the healing of microcracks in concrete, there was a marked increase in the compressive and bending strengths (figures 10 and 11). The distribution of calcium carbonate as a result of the self-healing of microbial concrete was studied[108] in detail, revealing carbonate deposits along the x-axis crack direction. The calcium carbonate content decreased away from the crack surfaces in the x-axis direction. There was a gradual accumulation of calcium carbonate in the y-axis direction. Various repair points gradually became connected, and the entire crack system was eventually filled completely. In the z-axis direction, calcium was deposited on the surface in the fracture direction when the crack was filled at the surface. The so-called hypoxia existing in the depths of cracks led to a meagre production of calcium carbonate.

Table 5. Apparent chloride diffusivity in various concrete specimens

Specimen	Exposure (days)	D_{app} (m²/s)
uncracked	49	5.634×10^{-12}
uncracked	133	2.704×10^{-12}
standard crack	49	5.161×10^{-12}
standard crack	133	2.597×10^{-12}
standard crack healed by high-viscosity polyurethane	49	4.347×10^{-12}
standard crack healed by high-viscosity polyurethane	133	2.062×10^{-12}
standard crack healed by low-viscosity polyurethane	49	3.442×10^{-12}
standard crack healed by low-viscosity polyurethane	133	1.920×10^{-12}

The self-healing behaviour of young concretes containing a 4wt% crystalline admixture was monitored[109] by measuring the permeability of cracked specimens and the crack widths. Three exposure conditions were used: water immersion at 15C or 30C, and wet/dry cycling. The specimens were pre-cracked after 48h, yielding crack widths ranging from 0.10 to 0.40mm. There was almost perfect healing of specimens which were held under water at 30C. It was inferior for specimens held under water at 15C, and was insufficient in the case of wet/dry exposure. Capsules of changing brittleness were developed[110] such that a high initial flexibility was imparted by adding a plasticizer to an ethyl cellulose matrix. During the hardening of concrete, the plasticizing agent was expected to leach out into the moist environment, thus leaving behind more brittle capsules which would break when a crack appeared. There was some incompatibility between the capsule wall and the polymeric healing agent within. The capsules became insufficiently brittle and the strength of the bond with the concrete matrix was too weak. Multilayer capsules were consequently introduced which had an impact resistance that was sufficient to survive concrete-mixing but could still break when a crack formed.

Encapsulated magnesium oxide, bentonite and quicklime powders have been used[111] for self-healing, with concentric glass macrocapsules enveloping swelling minerals (outer capsule) and water (inner capsule). Samples which contained concentric macrocapsules having various mineral combinations were cracked and healed under 3 curing conditions: ambient, high humidity and water immersion. The self-healing was monitored visually, and by measuring mechanical strengths and durability. Immersion in water produced the optimum healing efficiency, with some 95% of crack sealing and a strength recovery of about 25% within 28 days. Over 56 days there was an increasing trend in both crack-sealing and load-recovery. The healing kinetics indicated that the expanding minerals were hydrated during the initial healing period, and then slowly carbonated over time until the peripheral crack zone became sufficiently water-tight. Potential self-healing agents were impregnated into lightweight aggregates that were then mixed into concrete[112]. As an example, lightweight aggregates having diameters ranging from 4 to 8mm were impregnated with sodium silicate solution. Concrete specimens which contained the impregnated aggregates were then pre-cracked in order to create crack widths of up to 300µm after 7 days. Following 28 days of healing in water, specimens which contained the impregnated aggregates exhibited a recovery of some 80% of the pre-cracked strength. This was more than 5 times the recovery of control specimens. Specimens which were healed by the impregnated aggregates exhibited a 50% reduction in the sorptivity index, as compared with cracked control samples. This was very similar to the response of uncracked control specimens. The contribution made by sodium silicate to the production of more calcium silicate hydrate gel was confirmed. A review of the topic[113] suggested that 8 key factors governed the effectiveness of self-healing by encapsulation. These were the survival of mixing, the probability of a crack encountering a capsule, curing times and conditions, the effect of empty capsules upon concrete strength, the controllability of the release of the healing agent, the stability of the healing agent, the sealing ability and consequent recovery of durability and strength, and the repeatability of the self-healing action.

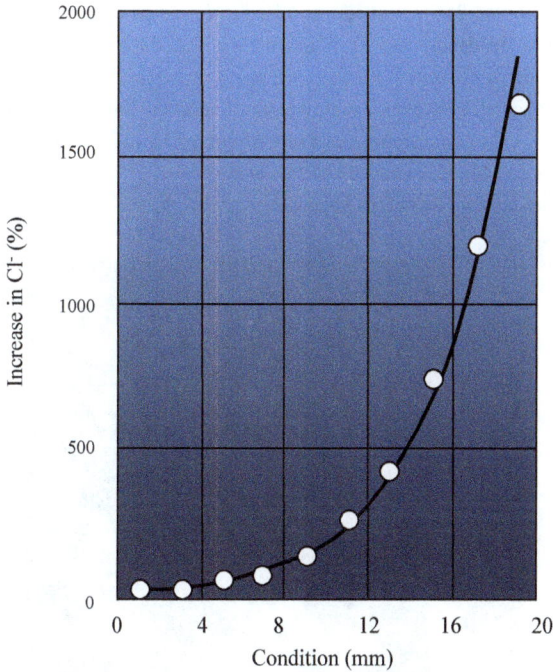

Figure 12. Cracked-induced increase in chloride concentration with distance from the exposed surface

Predictions[114] of the service life of self-healing concrete were based upon chloride-diffusion studies (figure 12). Cracks (300μm) have a large effect upon the penetration of chlorides in concrete. After 49 days of exposure, there was an increase in the chloride content at all depths in the concrete due to such a crack. The effect of the crack is quite limited close to the exposed surface, but increases almost exponentially (table 5) with depth in the concrete. The self-healing of cracks by encapsulated polyurethane constituted a partial barrier to the immediate entry of chlorides via cracks. Self-healing concrete was able to reduce the chloride concentration in a

Materials Research Forum LLC

https://doi.org/10.21741/9781644901373

crack zone by at least 75%. The service life of steel-reinforced self-healing concrete slabs in a marine environment could therefore be 60 to 94 years, while normal cracked concrete was expected to survive for only 7 years. Engineered cementitious composite exhibits a strain capacity of 4.4% and a deflection capacity of 7.9mm under tension and bending, thus avoiding the brittleness of normal concrete. The flexural and compressive strengths of cementitious composite are 12.2 and 45.8MPa, respectively. Under restrained shrinkage, the composite also exhibits a very low tendency to suffer fracture failure. Most relevantly, self-healing is observed[115] in engineered cementitious composite and its stiffness, tensile strain, tensile strength and resonant frequency exhibit a very high degree of recovery following self-healing and the properties approach those of virgin cementitious composite of the same age. The water permeability coefficient of pre-damaged composite decreases gradually with self-healing age, and eventually comes close to that of undamaged specimens.

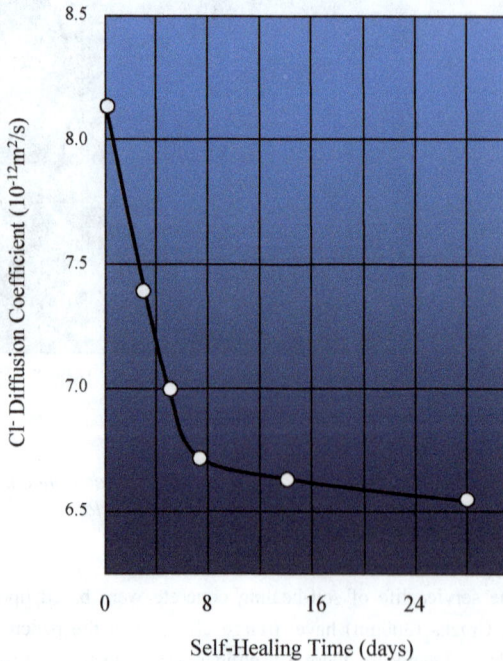

Figure 13. Chloride diffusivity as a function of healing time

The self-healing effect, upon reinforced and non-reinforced concrete beams, of microcapsules filled with calcium nitrate was again investigated[116] by comparing the initial stiffness, peak strength and deformation values with post-healing data. Crack monitoring was also used to evaluate crack healing as a function of time. It was found that the air content in samples which contained microcapsules was twice that in control samples. The addition of microcapsules lowered the flexural strength of concrete beams. Stiffness recovery was noted for samples with and without microcapsules and with or without steel reinforcement. Control samples exhibited the poorest stiffness recovery. The use of steel and microcapsules imparted better healing and improved the stiffness recovery by 38%. Image analysis showed that crack widths did not completely heal in control samples, while the use of microcapsules allowed the cracks to heal more efficiently. The best performance was that of microcapsule- and steel-containing samples, which exhibited 100% crack healing. The effect of urea-formaldehyde/epoxy microcapsules upon the mechanical properties of specimens was investigated[117], and the self-healing ability was judged by the crack-healing effect and chloride permeability. The results showed that the chloride diffusion coefficient decreased from 8.15 x 10^{-12} to 6.53 x $10^{-12} m^2/s$ after 28 days of healing (figure 13). The healing efficiency of the permeability increased to a maximum of 19.8%. The number of microcapsules, the particle size, the curing temperature and the degree of pre-loading had marked effects upon the self-healing efficiency. Urea-formaldehyde microcapsules with *in situ* polymerization and controlled particle-size distributions have also been used to improve the self-healing of cemented coral sand[118]. The particle-size distribution of the microcapsules was varied by using differing rotation-rates during synthesis. The particle-sizes of the microcapsules ranged from 0 to 650μm when the rotation-rate was varied from 150 to 800rad/min, and there was a linear relationship between the medium particle-size and the rotation-rate. For a given rate, the particle sizes had a normal distribution. The optimum particle-size distribution and volume fraction corresponded to 450rpm and 3%. The compressive strength of specimens was increased by about 85% when 3% of microcapsules were added. The particle-size distribution of microcapsules at 450rpm matched well with the pore-distribution in the specimen. Microcapsules having a size greater than 18μm were found to be broken after compression. The self-healing efficiency of microcapsules which were prepared at 450rad/min, with a volume fraction of 3%, was about 75%. The microcapsules could be triggered by cracking or by compression of the specimen and thereby self-heal internal cracks.

In order to observe the self-healing mechanism in concrete by using ultrasonic methods, cracks were introduced by 3-point bending[119]. The self-healing agent was able to improve the properties even after damage, and the main self-healing mechanism in this case was the reaction of sodium silicate with calcium hydroxide in the concrete. Concrete formulations were developed[120] which incorporated a self-healing system that was based upon silica microcapsules which contained an epoxy sealing compound and amine-functionalised silica nanoparticles. The sealing compound was integrated into the cementitious matrix, but the microcapsule content led to impairment of the mechanical behaviour. The inclusion of the additions also refined the pore distribution and increased the durability in aggressive media. An autonomous self-healing ability was noted which was surprisingly higher in samples having a lower content of healing additions. This ability depended upon the crack width and the healing period. The strength and fracture toughness of the glass/concrete interfaces often present in encapsulation-based self-healing concrete were assessed[121]. Uniaxial tensile tests were used to assess the bonding strength, and 4-point bend tests furnished the interfacial energy. Parallel numerical models were based upon the finite-element method. A cohesive-zone model was used to study the interface strength, and a virtual crack closure model was used to analyze the interfacial toughness. It was found that a glass/concrete interface could exhibit a maximum strength of about $1N/mm^2$, with a fracture energy of $0.011J/m^2$.

The possibility of using self-healing coatings to protect steel-reinforced concrete was investigated[122], with the primary aim being to withstand construction-site damage. When such damage was caused to conventional coatings, it impaired the ability of the coating to resist corrosion, given that the coating consisted of epoxy containing 10wt% of micro-encapsulated tung oil as the healing agent. In accelerated corrosion testing, the time-to-failure of steel-reinforced concrete with self-healing coatings was at least 3 times longer than that of steel-reinforced concrete with conventional coatings. Following 150 days of accelerated corrosion-testing, 83% of the samples with self-healing coatings suffered no corrosion. There was no difference between undamaged and deliberately damaged specimens, possibly because the damaged area was too small. It was suggested that self-healing coatings involving natural products were able to passivate reinforcement-bar surfaces when corrosion began, thus markedly increasing the protection against corrosion. Pull-out tests showed that self-healing coatings exhibited comparable bond stresses to those of conventional coatings. A self-healing system was described[123,124] which was based upon sodium nitrite and ethyl cellulose or sodium monofluorophosphate and ethyl cellulose microcapsules. The sodium salt acted as the healing agent and the ethyl cellulose was the container. The average size of the former microcapsules was 270μm, while that of the others was 410μm. The chosen capsules were mixed into the concrete, making up 5wt% of the cement, and corrosion-acceleration tests were carried out by wet-dry cycling at 50C. Pitting corrosion of a control sample occurred after 72h of wet-dry cycling, while samples which contained the sodium nitrite microcapsules were still smooth after 576h. The samples with sodium monofluorophosphate microcapsules began to rust after 432h. This demonstrated that the self-healing systems possessed marked anti-corrosion properties.

Healing agents which are embedded in concrete travel to the damage zones mainly via capillary flow through cracks. A model system[125], for studying the capillary flow of low-viscosity cyanoacrylate and ground blast furnace slag in a water suspension, employed glass capillaries and channels which were formed from a range of concrete mixes. The healing agent closely obeyed Poiseuille's law, and exhibited increases in viscosity during the 40 minutes that it was exposed to a cementitious environment. Numerical simulations of the capillary rise of healing agents in a discrete crack confirmed that the rate of damage and degree of saturation of the concrete should have a large influence upon the choice of healing agent. The durability of a self-healing concrete in an aggressive environment was simulated[126] by employing freeze-thaw cycles and salt-spray tests. The ultra-high strength reinforced concrete incorporated an autonomous self-healing system which was based upon the reaction of an epoxy compound that was enclosed within silica microcapsules, and amine-functionalised silica nanoparticles which were distributed within the cementitious matrix. The effect of the aggressive environments was analysed for crack widths of 150 and 300μm. Capillary water absorption tests, and electron microscopy, confirmed an increased durability of the autonomously self-healed material during freeze-thaw and salt-spray testing; as compared with that of a control sample.

Calcium nitrate tetrahydrate was again evaluated[127] as a healing agent in concrete, especially with regard to improving the mechanical properties by altering the micro-encapsulation process and modifying the concrete formulation. This showed that the introduction of newly-developed microcapsules and of a de-foamer into the formulation were key factors in minimizing the air content of microcapsule-containing concrete. They resulted in improved mechanical properties as compared with those of controls. Differences in the modulus of elasticity of control samples and microcapsule-containing samples were minimal, although microcapsule-containing specimens having the lowest concentrations were much less affected by damage cycles than were the control samples. Some microcapsule-containing specimens offered a marked healing potential with regard to the compressive strength and modulus of elasticity following damage. The optimum microcapsule concentration was found to be between 0.25 and 0.50wt% of cement under axial compression. The use of silica gel as a self-healing agent was again investigated[128]. It was added, in concentrations of 0.1, 0.2, 0.3 or 0.4wt%, to the mix and the concrete was subjected to curing periods of 7, 14 or 28 days. The compressive strength of concrete with 0.3% of silica gel was 6.54% greater than that of control specimens after 28 days. This was deemed to be the optimum content.

The self-healing behaviour of concrete which incorporated re-cycled aggregate was investigated[129]. The re-cycled mixtures involved replacing natural aggregates with 0, 30 or 100% of re-cycled concrete gravel and 30% of re-cycled concrete sand. The ratio of water to binder was kept constant, and the re-cycled concrete aggregate was initially in a saturated surface dried state. The long-term maximum variation in drying shrinkage, due to the re-cycled concrete gravel, did not exceed 15%. The re-cycled concrete aggregates were affected by drying as soon as the concrete was subjected to desiccation. A close correlation was observed between the 1-day compressive strength and the 18-month carbonation depth, and the relative recovery of cracks reached up to 60%. The self-healing performance of calcium nitrate microcapsules was studied[130] in concrete beams which were reinforced with steel and shape memory alloys. The results suggested that the microcapsules had a direct effect upon decreasing the flexural strength, and this was attributed to the presence of a high air content when microcapsules were present. Although microcapsules tended to lower the flexural strength, stiffness recovery was greater for specimens which contained microcapsules. Observations which were made following water curing indicated that specimens with microcapsules exhibited the best crack-healing efficiency, regardless of the use of steel or shape memory alloys.

So-called heavyweight concrete, made from aggregate with an oven-dry density greater than $3.5g/cm^3$, is commonly used to contain radioactive waste. It is thus required to have a high structural integrity and a high resistance to moisture infiltration. The self-healing of heavyweight concrete has been tackled[131] by using an expansive additive, fly-ash and organic fibers. This led to an early reduction in water permeability, but the internal crack width hardly decreased and crack closure was observed only in the surface region. The addition of the expansive additive led to self-healing in the early stages, and this was attributed to chemical pre-stressing and to the formation of ettringite which filled fine cracks. The inclusion of fly-ash and organic fibers appeared to have a long-term self-healing effect. In the presence of water, cracks in reinforced concrete structures can generally heal naturally or with the help of additives. When the water permeability and self-healing behaviour of a high-performance concrete with fibers, or with fibers plus a crystalline admixture, was studied[132] it was found that, under monotonic loading conditions, both types of sample had maximum crack widths which were 39% smaller, and water permeabilities which were 3.1 times poorer than those of the high-performance concrete. During 7 days of constant loading in a continuous water-flow, cracks in both samples healed completely, as compared with only 60% in the case of the basic high-performance concrete. The self-healing was slower for the crystalline-admixture samples than for the fiber samples, but a higher load had to be applied to the former in order to regain the initial permeability. The self-healed fiber samples had calcite and ettringite in the cracks, while the additive samples had aragonite. A comparison was made[133] of the effect of 5 or 10wt% of magnesium oxide expansive agent upon the self-healing behaviour. For cracks having a given initial width, the higher the reactivity and the greater the magnesium oxide content, the higher was the restrained expansion rate and the better was the self-healing response. The magnesium oxide had a better self-healing effect upon new cracks with initial widths of less than 0.40mm. The effect markedly weakened when the initial width extended beyond 0.40mm. The 7 and 28 day compressive-strength recovery-rate of specimens with magnesium oxide was improved by 55 to 68% when compared with a control. Self-repair experiments were carried out[134] by using fiber-glass tubes as adhesive carriers. The flexural strength following repair, and the initial flexural strength of the control, were used to characterize the repair performance. The repair effect increased as the environmental temperature increased; the performance at 30C being 13.3% better than that at -15C. The smaller the initial crack width, the better was the repair effect. Microcracks with an initial width of less than 1.0mm could be repaired. When a fiber-glass tube was placed in a form of a trapezoid, the performance was better than when it was a rhombus. The tube placement played a predominant role in self-healing when the initial crack width was less than about 2.0mm. The fracture of a capsule under uniaxial tension in an encapsulation-based self-healing concrete system was studied[135] for the case of a circular capsule embedded in mortar. Zero-thickness elements were used to represent potential cracks, and the effects of the fracture strength around the interfacial transition zone of the capsule were analyzed. Numerical results indicated that an increase in the strength of the interfacial transition zone around the capsule could increase the load-carrying capacity of self-healing concrete. For a given fracture strength around the capsule interface, the probability of fracture of the capsule depended markedly upon the core/shell thickness ratio.

A study was made[136] of the self-healing of reactive-powder concrete which was modified with nanofillers. The filler was nano-SiO_2, nano-TiO_2 or nano-ZrO_2, and samples were subjected to water-curing or air-curing. All of the fillers could improve the self-healing performance of the reactive-powder concrete, but the improvement was clearer when the concrete was cured in water. Following water-curing, samples which involved nano-SiO_2 exhibited the greatest self-healing ratio with regard to the compressive and flexural strengths: 1.31 and 1.19, respectively. These represented corresponding increases of 39.4 and 33.7% over those for reactive-powder concrete without nanofillers. The effect of a nanofiller was attributed to the provision of a hydrating environment for unhydrated cement particles and of nucleation sites for hydration products. It also improved the 3-dimensional network structure of the concrete matrix, produced more fine cracks and diversified the crack-propagation direction. Pozzolanic reaction of these nanofillers, especially nano-SiO_2, led to additional calcium silicate hydration. Self-healing was studied[137] in high-cement, low water/binder ratio, composites which were aged for 2 or 20 months. Mechanically-induced cracks which were up to 800μm wide were analysed. The maximum width of the healed cracks was 460 and 388μm, respectively. The healing process was affected by crushed material in the cracks. The results indicated that a dense cement matrix could reduce the self-healing efficiency. The over-grinding of Portland cement causes excessive shrinkage, and impairs the self-healing ability of concrete. In the case of Portland cement which contained more particles in the 30 to 60μm size range, the concrete exhibited[138] a superior self-healing ability after 28 days. The autogenous self-healing of the concrete was affected by continued hydration and carbonation. After 7 and 30 days, the autogenous self-healing was controlled mainly by continued hydration and carbonation. The particle size could influence the hydration by changing the non-hydrated cement content, and influence the carbonation by changing the $Ca(OH)_2$ content. It could be concluded that a suitable distribution of cement particle sizes, with its associated content of $Ca(OH)_2$ and un-hydrated cement, could improve the self-healing ability.

Microcapsules having toluene-di-isocyanate as the healing agent and paraffin as the container were used[139] for the self-healing of mortars. The preparation-temperature, agitation rate and paraffin/contents mass ratio had a marked effect upon the core fraction of the microcapsules. The optimum microcapsules had a paraffin/contents mass ratio of 1:2. The core fraction of microcapsules having the optimum parameters was 66.5%, and the particle size ranged from 30 to 300μm with a peak at 90μm. The microcapsules were regular spheres and the shell thickness was equal to about 10% of the diameter. As compared with a control sample, the compressive strength of mortar with 3wt% of microcapsules was increased by 28.2%. The retained ratio of compressive strength was 77.2% following 48h of self-healing. Cracks having a width of less than 0.4mm were self-healed within 6h.

The flexural strength and stiffness of a notched bend specimen, before and after healing, were compared[140] and used for self-healing assessment. The neutral axis of the specimen was measured by using strain gauges which were attached to the crack-propagation region. The strength and stiffness of the concrete recovered following healing, but the change in the location of the neutral axis, before and after healing, was insignificant. The neutral axis moved slightly upwards during crack propagation, but there was an insignificant movement when the crack was closed or re-opened. It was found that, the longer the healing period, the greater was the healing effect. For a given healing period, the later that a crack was introduced the less effective was the healing. It appeared that the curable area was limited to that portion between the neutral axis and the crack tip, rather than covering the opened crack area.

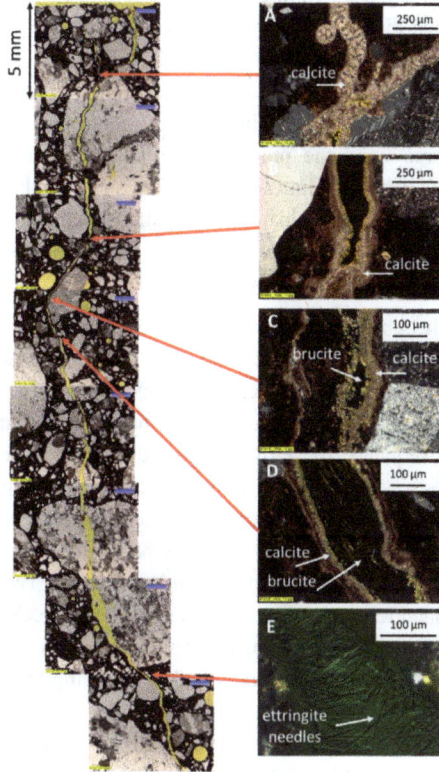

Figure 14. Overview image of an epoxy-impregnated crack with a depth of 40mm, showing the variation in self-healing product. Reproduced from Mineralogical Sequence of Self-Healing Products in Cracked Marine Concrete, Danner, T., Jakobsen, U.H., Geiker, M.R., Minerals, 9[5] 2019, 284 under Creative Commons Licence 4.0 (see reference list for full DOI).

An active crack-width technique has been developed[141] in order to reduce variations in the width within an experimental series and thus guarantee consistent permeability results. This allowed it to be confirmed that the main factors which contribute to permeability variation are the mean crack width at the crack mouth and the internal crack geometry. Variability in the mean crack width could introduce a threefold higher variation in permeability data. Unlike the crack width, the internal geometry of the crack cannot be determined directly but can cause permeability differences of more than 25% in specimens having an identical surface crack-width. The conclusion is that, even if crack widths can be closely controlled, the main source of variability in permeability is the internal geometry; which cannot be similarly controlled. A study[142] of the effect of crystalline admixtures upon the properties of concrete samples made using ordinary Portland cement or Portland limestone cement showed that the water-permeability coefficient was decreased threefold while the self-healing ratio was increased. The self-healing test data indicated that there was a more rapid sealing of crack-widths of up to 250µm in specimens which contained a crystalline admixture.

The extent of self-healing of cracked concrete beams after 25 years of marine exposure was investigated[143]. Varying the silica fume (4 or 12%) or fly-ash content (0 or 20%) had no effect upon the nature of the self-healing products or the extent of self-healing. Crack-widths that were smaller than 0.2mm appeared to be closed and, with increasing crack depth, there was a sequence of self-healing products. At up to 5mm from the exterior surface, only calcite was precipitated. This was followed by brucite layers at depths of 5 to 30mm, with an occasionally fraction of calcite. Ettringite alone was found at depths that were greater than 30mm (figure 14). This particular sequence was attributed to an increasing pH, with depth, of the solution within the crack. The self-healing of cracks in marine-exposed concrete was explained in terms of the precipitation of ions from sea-water, due to reaction with ions from the cement paste in the outer part of the crack, together with the dissolution and re-precipitation of ettringite at greater depths. Cracking under a splitting tensile loading, and the morphological characteristics of cracks, were investigated[144] by using disc specimens. Tensile stress versus crack opening displacement curves during loading and unloading were analyzed for various unloading displacements. The results indicated that cracking led to a decrease in the stiffness and, the greater the crack opening displacement, the more the stiffness decreased. Upon re-loading healed disk specimens, the bond strength between any self-healing products and the concrete matrix was relatively weak. Although the stiffness was partially restored, the degree of recovery was relatively low when compared with the initial stiffness. The self-healing mechanism consisted of continued cement hydration, crystallization and blockage of loose particles.

Microcapsule-based self-healing concrete was investigated[145] by using laboratory and field tests. The self-healing behaviour was evaluated by using compressive strength tests and a rapid chloride migration test. This indicated that the self-healing of concrete which contained 10% of microcapsules gradually increased with time. Inclusion of the microcapsules led to a marked increase in long-term shrinkage, but that degree of shrinkage was acceptable for practical use. No appreciable differences in strain were found between experimental and field tests. Autonomous crack-healing by encapsulated polyurethane was explored, as a means for reducing the perpendicular-to-crack entry of chloride, by tracking the chloride penetration front using $AgNO_3$ and electron probe micro-analysis[146]. The healing mechanism was effective in reducing the chloride concentration in the direct vicinity of the crack, and caused a reduction in the perpendicular-to-crack chloride penetration. A general test method has been proposed[147] for the evaluation of the self-healing ability of cement-based materials in terms of their resistance to chloride penetration. Steady-state chloride migration has been used to measure the diffusion coefficients of cracked mortar specimens which contain crystalline, expansive and swelling admixtures. The results showed that the time which was required to reach a quasi-steady state decreased, and the diffusion coefficients increased, as the potential increased. This was because of the potential drop within the migration cell and the self-healing which occurred during the test. The use of a high potential was advised in order to minimize the test duration, provided that the temperature did not rise excessively during the test. An index of self-healing capacity could be based upon the rate of charged chloride-ion passage through a crack.

In a recent study[148], a light-induced self-healing mechanism was developed, for the automatic healing of cracks in concrete, in which cementitious capsules were filled with an ultra-violet adhesive and an ultra-violet light-emitter. The capsule-shell was made up of three separate but bonded layers. The capsules were able to survive within the cementitious material and yet be broken upon crack formation. Bacterial self-healing concrete was cast into the form of a slab[149]. The bacterial additive was made from a mixed ureolytic culture, mixed with anaerobic granular bacteria. Cracks at the bottom of a specimen which was subjected to wet/dry cycles exhibited the most crack closure. The sealing efficiency of cracked specimens which were submerged in water for 27 weeks was at least 90%.

Finally, and very recently, a new form of biomimetic cementitious material[150] comprised 3-dimensionally printed tetrahedral mini-vascular networks which stored, and then delivered, a healing agent to damage sites within a cementitious matrix. The vascular network served firstly to preserve the healing agent and allow it to survive the mixing process. It then released the healing agent when the cementitious matrix was damaged, with a minimal effect upon the mechanical properties of the matrix. Mini-vascular networks which were filled with sodium silicate, and embedded in concrete, were able to act as healing-agent reservoirs that were available for multiple-damage healing. The healing agents, encapsulated within the networks, could be transported to crack zones in concrete via capillary action; thus recovering the strength, stiffness and fracture energy. A biomimetic 3-dimensional vascular network was designed[151] according to Murray's law of circulatory-blood volume-transfer. The structures were constructed by 3-dimensional printing and incorporated into a cement-based matrix. One-dimensional and 2-dimensional models were also embedded into cement prisms for comparison purposes. Load-recovery was used to monitor the recovery of mechanical properties when a sample was cracked and pumped with sodium silicate for 28 days. This proof-of-concept study confirmed the ability of all of the vascular systems to deliver a healing agent following a damage event, with the 3-dimensional system in particular exhibiting a markedly superior healing ability. In another study[152] tests were performed on a pressurised vascular cementitious material system in which the healing-agent was a low-viscosity cyanoacrylate. One set of tests involved plain concrete notched prismatic beams and examined the mechanical response as a function of the healing period, rate of loading and the healing-agent pressure. Another set involved direct tension testing of doubly-notched prismatic specimens, having differing crack opening displacements during the healing period. Healing was here allowed to take place in cracks that were held stationary for a time, with the degree of mechanical recovery being measured for various healing periods. A simplified damage-healing model was used to interpret the results. A study was made[153] of the design of vascular tubes, for enhancing healing behaviour and recovering mechanical properties, by reinforcing hybrid resin with phenol-treated carbon fibers and inserting macrovascular tubes. The effect of the phenol treatment was to increase the tensile strength of composites by 7%. A model material system was studied which comprised cyanoacrylate-filled channels in concrete structural elements[154]. It was used to consider the transport and curing properties of the healing-agent in terms of the capillary-flow behaviour of cyanoacrylate in a static natural crack, of the sorption of healing-agent through a cracked surface and into a concrete specimen, of the development and progress of a cyanoacrylate curing front next to a concrete substrate and of the dynamic flow characteristics of cyanoacrylate in capillary channels. Theoretical relationships were established for each of these processes, and these could be used to help to characterise the behaviour of the material. In each case, the processes exhibited an appreciable degree of variability but also strong behavioural trends.

These various healing strategies will now be reviewed in more detail.

Polymers

The self-healing of concrete has long been observed in traditional concrete use, but this occurred mainly due to the behaviour of unhydrated cement particles when exposed to water. So-called super-absorbent polymers exhibit phenomenal moisture uptakes of more than 90% of the original polymer weight, making them useful water reservoir candidates in concrete-related applications. The use of encapsulated polymers in concrete, where the capsules and/or contents can be polymers, is instead aimed mainly at targeting specific cracks during fracture. Self-healing aims at the autonomous healing of small cracks having widths of the order of a few hundreds of micrometers. Natural crack healing relies upon the appearance of $CaCO_3$ and various minerals in cracks, a phenomenon which is best suited to the healing of static cracks. The addition of super-absorbent polymers to cementitious mixtures promotes self-healing of the material[155]. When cracks occur, polymer within the crack swells upon contact with water and subsequently releases the water, which then stimulates further hydration of any unhydrated cement particles and promotes calcium carbonate crystallization. On the other hand, the presence of super-absorbent polymer affects the mechanical behaviour of the cementitious material by creating macro-pores. In order to counteract the reduction in strength, part of the cement can be replaced by nanosilica. In samples which contained super-absorbent polymer, an immediate sealing effect was observed, together with crack closure. A lesser effect upon the mechanical properties, and good healing, was observed in mixtures which contained both nanosilica and super-absorbent polymer. The above macrovoids which resulted from the absorption/desorption of super-absorbent polymers could markedly reduce the mechanical properties[156]. Super-absorbent sodium alginate hydrogels were prepared by cross-linking Ca^{2+} and sodium alginate. At lower Ca^{2+} concentrations, higher swelling potentials were observed; consistent with the formation of a less dense network. This dependence upon Ca^{2+} concentration probably avoided the formation of large macro-pores during cement mixing and led to increased swelling when cracks occurred and water entered cracks. Compressive strength measurements showed that the addition of 0.5wt% of sodium alginate led to a negligible (0.014%) reduction in compressive strength, due to the limited swelling capacity in pore solution. In a sodium silicate self-healing system, sodium alginate also accelerated the precipitation of calcium silicate hydrates by providing Ca^{2+} ions which made the healing process more efficient.

In one test, plates which contained various amounts of super-absorbent polymer were impacted[157] after 1 week, 1 month, 4 months or 6 months and were stored under conditions of wet/dry cycling and a relative humidity of 95. After 28 days of healing, the plates were impacted and again healed. The specimens which contained super-absorbent polymer exhibited a more ductile behaviour during impact loading, as compared with control samples. Even after the second impact loading of healed samples, superior healing was caused by wet/dry cycling as compared with healing at 95% relative humidity. The assessment of cement-based composites which included super-absorbent polymer[158] showed that the flexural strength was increased by 78%. The self-healing/sealing behaviour was assessed, showing that cement/polymer composites which contained 2% of super-absorbent polymer per weight of cement exhibited a self-healing/sealing ability when an artificial crack was introduced. Such a crack was completely healed over most of its length following 28 days of treatment. Super-absorbent polymers have been used[159], together with various superplasticizer contents to produce self-healing concrete. A special mould was used to create controlled crack-patterns throughout the entire volume of a sample. Experimental results showed that cracks with widths of up to 234µm were healed when 2.2% of superplasticizer was added per weight of cement.

Encapsulation-based self-healing[160] is initiated by the occurrence of a crack and continues with the chemical reaction of self-healing agents, contained in capsules, which are released into the crack. Polyurethane-based healing agents, contained in glass or ceramic tubes, are commonly used in self-healing cementitious materials. Mechanical strength and durability are the most widely used criteria for self-healing efficiency assessment. The first consideration in planned self-healing is the creation of capsules which are able to survive concrete-mixing, and yet nevertheless break when cracks appear. Three polymers having a low glass transition temperature have been considered: poly(lactic acid) (transition at 59C), polystyrene (transition at 102C) and polymethylmethacrylate/n-butylmethacrylate (transition at 59C). Heating the capsules, in order to convert them from brittle to rubbery, greatly increased their survival[161]. Part of the capsule, which had previously survived concrete-mixing, still broke upon crack formation. With the same philosophy, 300 to 700µm microcapsules which contained sodium silicate solution were made[162] from a cross-linked gelatin/acacia-gum material. The thickness of the microcapsules ranged from 5 to 20µm. This shell material possessed switchable mechanical properties in that, when it was hydrated, it was rubbery but, when dry, was glassy. These microcapsules survived drying and hydrating cycles and maintained their structural integrity when exposed to solutions which simulated the strongly alkaline concrete environment. They were also very stable at up to 190C. The microcapsules survived being mixed with cement, yet ruptured upon crack formation and successfully released the sodium silicate solution.

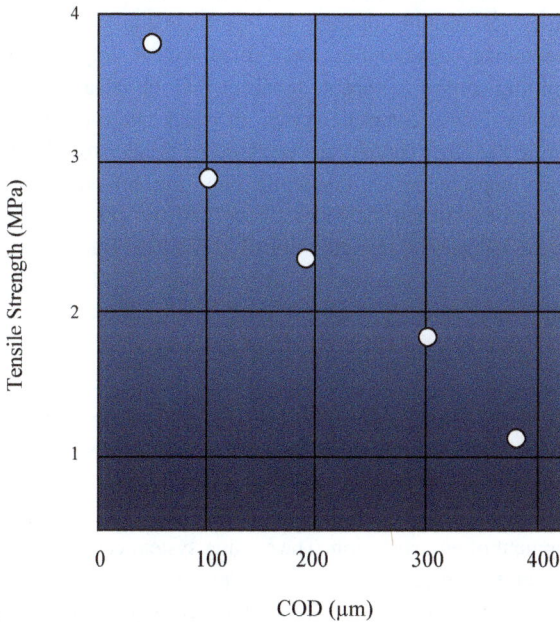

Figure 15. Failure stress of hardened polyurethane resin in
concrete as a function of crack opening displacement

There is also the question of the ability of flexible polymers to bridge healed cracks and heal moving cracks. Fundamental tests were performed[163] on the release and curing of encapsulated polymer precursors which offered a wide range of properties, including mechanical strength following hardening. The results permitted the identification of precursors which guaranteed good crack-filling, and thus the successful sealing of cracks. Although such sealing was achieved, the recovery of mechanical stiffness was limited to 30%; and that only for crack-opening displacements of up to 20μm. Polymeric capsules have an advantage over glass capsules in that they are easier to prepare and offer a wider range of mechanical properties; the latter factor making it possible to guarantee capsule rupture and healing-agent release when intercepted by a crack of a certain size. Polymethylmethacrylate, for example, ruptures at an average crack size of 69 and 128μm for a wall-thickness of 0.3 and 0.7mm, respectively[164]. Thicker walls rupture at crack sizes which are much greater than 100μm. An elongation at fracture of 1.5% is recommended for polymers.

Experimental and numerical methods have been used[165] to study crack-filling in encapsulation-based self-healing concrete. The specimen consisted of two small concrete blocks which contained capsules that were filled with a polyurethane-based healing agent, thus permitting control of capsule breakage and the release of the healing agent. One arrangement involved a 2-capsule system in which one capsule contained the pre-polymer fluid while the other contained a water-based accelerator. Another arrangement involved a single capsule which contained only the pre-polymer fluid. A linear relationship was found between the failure stress and the crack opening displacement (figure 15), with the failure stress ranging from 3.7MPa for a crack opening displacement of 50.8mm to 1.2MPa for a crack opening displacement of 381mm. This represents the mechanical contribution made by a polyurethane resin released into a crack within a concrete structure and gives some estimate of the degree of recovery of mechanical properties.

As noted elsewhere, the autonomous healing of cracks by encapsulated polyurethane has been investigated as a means of reducing the corrosion of concrete reinforcements. Such beams were exposed weekly to a chloride solution and various electrochemical parameters were measured[166] in order to monitor the effect of the self-healing mechanism upon the development of any corrosion. The healing system consisted of brittle capsules, filled with the agent, that were embedded in the concrete. The capsules were 50mm-long sections which were cut from borosilicate glass tubes having an internal diameter of 3mm and a wall thickness of 0.175mm. The healing agent was either a non-commercial polyurethane precursor with a viscosity of 6700mPas at 25C, or a commercially available polyurethane precursor with a viscosity of 200mPas at 25C. The ends of the capsule were sealed with polymethylmethacrylate. In the case of cracked beams, corrosion was detected within a space of 10 weeks, with visible pitting being present on the reinforcing bars. The use of low-viscosity polyurethane could markedly reduce the degree of corrosion which was present during the propagation stage and, in this case, the reinforcing bars exhibited no visible damage. Crack-healing with the high-viscosity polyurethane appeared to be unsuccessful in blocking the entry of chloride, via cracks, due to incomplete crack-filling. Although the entry of chloride was reduced, the crack-healing could not prevent the initiation of corrosion. The specimens exhibited superior behaviour during propagation. Even after 14 weeks of exposure, the driving potential was comparable to that found in uncracked specimens. The concrete resistance of low-viscosity specimens in the cracked region was comparable to the resistance of uncracked specimens during the entire period of exposure; again confirming that crack-healing was effective in blocking, at least partially, the entry of NaCl solution via cracks.

In another system, quartz-glass capsules which contained a one-component polyurethane, diluted with acetone, were embedded in the concrete[167]. The capsules were made from glass tubes having a length of 30mm, an internal diameter of 8mm and a wall-thickness of 1mm. A polyurethane mixture was then injected. The self-healing ability of the polyurethane was judged by performing 3-point bending tests. When the acetone/polyurethane mass ratio was 1:5, the viscosity and surface tension were low and the healing effect was at its strongest. The rate of strength recovery following 48h of manual crack-healing was 75%. When the capsules were embedded in concrete, the flexural strength was increased by 6 to 30%. The embedded capsules broke upon crack formation: the bond strength between the capsules and the concrete was about 0.63N/mm^2, which permitted fracture of the capsules due to crack formation. When capsules containing healant with an acetone/polyurethane mass ratio of 1:5 were embedded horizontally, this imparted the highest degree of self-healing of cracked specimens; corresponding to a healing efficiency of 67%.

The self-healing of engineered cementitious composites has recently been explored[168] by adding light-burned magnesium oxide and super-absorbent polymer in order to maintain the ductility, as measured by uniaxial tensile tests. The results indicated that the oxide was helpful with regard to preserving strength, but did not help with regard to ductility. The initial cracking strengths for the plain material, for samples which contained just the oxide, for samples which contained just the polymer and for samples which contained both were 2.38, 2.58, 2.23 and 2.43MPa, respectively. The polymer meanwhile had opposite effects: The corresponding ultimate tensile strains were 2.90, 2.53, 3.22 and 2.75%, respectively. The effects were directly related to changes in the oxide and polymer. The hydration of MgO produced $Mg(OH)_2$, and growth of the latter crystals led to expansion. This was helpful with regard to the strength, but not the ductility. The polymer is not initially active and cannot hydrate. It later swells to form gel-like structures but, when it releases water, it shrinks and leaves a honeycomb of voids which constitute internal defects that affect strength and ductility. That is, the polymer improves ductility but decreases strength. Changes in the healing ability were judged in terms of the relative seepage ratio. The water permeability decreased with increasing number of wet-dry curing cycles. This initially increased rapidly but then slowly stabilized; indicating that self-healing was occurring. Addition-free specimens required 10 to 12 cycles to self-heal. Those with oxide alone added required 8 to 9 cycles, those with polymer alone required 7 to 9 cycles and those with both required only 5 to 6 cycles. This represented a 50% reduction in healing time. The $Mg(OH)_2$ could cause micro-expansion during its hydration, thus tending squeeze cracks shut. The hydroxide could also fill the cracks. The polymer could instead store a large amount of water in the form of a gel and

Materials Research Forum LLC
https://doi.org/10.21741/9781644901373

then slowly release it. Under dry conditions, it could provide the water required for the hydration of the oxide and other components and thus accelerate the formation of hydration products. The joint effect of the oxide and the polymer was therefore synergistic; causing self-healing to be faster and more effective.

In similar work[169], fly-ash and a super-absorbent polymer were used to promote self-healing, with the latter being judged by gradual decreases in water discharge through a crack. Specimens were pre-loaded in order to generate a crack, and were then exposed to continuous water immersion or exposure to wet-dry conditions. The crack width decreased with increasing fly-ash replacement ratio, while increased additions of the polymer limited water discharge through the crack due to its swelling to form a gel. The combined effects could lead to 100% crack closure and the restoration of the original permeability following 28 days of healing. The crack closure was attributed to the development of self-healing products such as calcium carbonate and calcium silicate hydrate. The polymer was responsible for a slump reduction which resulted from the adhesive effect of the swollen material and the absorption of free water. The initial absorbent ability could be suppressed by adding concentrated Ca^{2+} to the mix because it bonded to carboxylic groups in the acrylate chains of the polymer and led to the desorption of stored liquid. The addition of fly-ash improved damage prevention, in that specimens which contained fly-ash developed finer cracks, under a given load, than did specimens without fly-ash. The use of fly-ash having a higher calcium content could also permit the use of cement formulations having a higher polymer replacement ratio. The pozzolanic reaction in fly-ash could also greatly improve the self-healing effect that resulted from the appearance of newly-formed calcium silicate hydrate products. The compressive strength tended to decrease with increasing polymer content when the latter was above 4wt% of cement. Specimens that were healed by water immersion exhibited greater self-healing than did those which were subjected to wet-dry cycles. This was because the presence of water increased the hydration of the fly-ash and other unhydrated components and consequently promoted self-healing. The initial water discharge was some 55% lower for polymer-containing samples than for polymer-free material. Crack-sealing of polymer-containing specimens was promoted by the formation of an impermeable gel which expanded to seal an opening crack as soon as it was exposed to water. Increasing the polymer replacement ratio improved crack-sealing, due to plentiful moisture supply by the polymer. The polymer also accelerated carbonation, which tended to promote self-healing, by producing a higher degree of calcium carbonate deposition along the open crack. Following 28 days of self-healing, those specimens having initial crack widths of less than 0.25mm were effectively closed, and healed specimens contained permanent hydrated deposits in the damaged area. Calcium carbonate was found in appreciable quantities around the outer region of healed specimens, while calcium-silicate hydrate largely concentrated around fly-ash particles which were at some depth. The self-healing ability of concrete was, as before, increased by the synergistic effect of fly-ash plus polymer.

In recent work[170], a polymethylmethacrylate nanocapsule adhesive system and mini-emulsion polymerization has been developed in which a resin and accelerator, and a hardener, are separately encapsulated in polymethylmethacrylate shells. This system was investigated by embedding them the shells in a mortar matrix. They survived mixing and hardening, and the alkaline environment. The stress fields which were associated with propagating cracks broke the shells of the core–shell structures and released the healing agents, which then quickly bridged nascent and early-stage fractures (less than 10μm). Long-term healing was assured by the formation of polymorphic calcite crystals in the presence of moisture and CO_2. This improved the durability of the mortar by filling gaps. Previous work[171] had considered 3 types of nano-capsule with regard to availability, strength and temperature-tolerance. These had included polyethylene as the shell and nano titanium oxide as the core, and polyethylene as the shell and nano silicon dioxide as the core. It was concluded that the optimum content, 0.5%, of SiO_2 nano-capsules led to a higher compressive strength. Previously investigated capsules had consisted[172] of cement or of a combination of cement and super-absorbent polymers, granulated by polyethylene glycol and followed by a waterproof layer of epoxy resin and sand. The core materials of the capsules could react with water, with dissolution of the polyethylene glycol and swelling of the super-absorbent polymers in cracks. Specimens which contained capsules exhibited a high sealing-ratio for cracks of less than 400μm. Cracks wider than 200μm could be bridged due to capsules with super-absorbent polymer. Water-tightness plus flexural and compressive strengths were recovered. Mortar mixes that contained polymer-coated pellets encapsulating various potential self-healing agents were investigated[173] in which the pellets replaced 10% of the sand by weight. Mortar prism specimens were then pre-cracked so as to produce a 0.3mm crack after 7 days. The addition of the coated pellets improved crack mouth and depth sealing by 20 to 60%, as compared with control samples. The water absorption and gas permeability of the healed material were satisfactorily low. The pellet-containing samples also exhibited a considerable recovery of flexural strength and stiffness following 2 cycles of water curing of cracked specimens. Self-healing cementitious composites were prepared[174] by mixing epoxy resin based microcapsules with cement. The strength, cohesion and internal friction angle of specimens gradually decreased with increasing microcapsule content or increased pre-loading stress, while the degree of deterioration gradually increased with increasing microcapsule content or pre-loading stress. The peak strain meanwhile increased to various degrees. The mechanical properties of healed samples were recovered to various degrees as compared with those of damaged samples. The microcapsules had a markedly positive effect upon the recovery-rates of mechanical properties. The healing effect increased with increasing microcapsule-content.

The ambient temperature has a marked effect upon the self-healing ability of microcapsules. The self-healing ability of microcapsules with various shell compositions (paraffin, paraffin/polyethylene wax, nano-SiO_2/paraffin/polyethylene wax), containing toluene-di-isocyanate, was investigated[175] at 10, 30, 50 and 60C. The self-healing ability of mortar containing various microcapsules increased from 10 to 50C. When the temperature reached 60C, the self-healing ability of mortar containing paraffin microcapsules clearly decreased while the self-healing of mortars containing paraffin/polyethylene wax microcapsules or nano-SiO_2/paraffin/polyethylene wax microcapsules was almost unchanged. Self-healing of the latter was better than that of the other samples at a given temperature. The compressive-strength recovery rate, chloride diffusion coefficient recovery-rate and harmful-pore ratios of nano-SiO_2/paraffin/polyethylene wax samples, pre-damaged at 50C and self-healed for 7 days, were 94.1, 81.2 and 47.3%, respectively. When the ambient temperature reached 50C, surface cracks of samples containing any type of microcapsule could be completely self-healed in 5h. When the temperature was 60C, the healing ratio of 0.4 to 0.5mm cracks in samples with paraffin capsules decreased to 49.52% after 6h. Surface cracks in samples with nano-SiO_2/paraffin/polyethylene wax capsules could completely self-heal in less than 3h.

The synthesis of inorganic shell microcapsules is very difficult, but epoxy resins have been encapsulated in silica shell microcapsules, with an average particle size greater than 100μm, by interfacial polymerization[176]. These silica-shell microcapsules exhibit a good thermal stability together with brittle rupture, a high Young's modulus and hardness of the shell and a satisfactory deformation resistance. The micromechanical properties were attributed to the regular tetrahedral crystal structure of the silica shell materials.

Self-healing microcapsules with fluorescein sodium labelled epoxy resin as a core material and urea-formaldehyde resin as the shell material were synthesized[177] by *in situ* polymerization and ultrasonic dispersion. The results showed that fully spherical particles with sizes in the range, 150 to 300μm, and having some surface roughness were well-dispersed. The addition of the fluorescein had no great effect upon the micromorphology, chemical structure, thermal stability or reactivity of the capsules. The point of adding the fluorescein was that, when a microcrack occurred at the surface of cement paste, the microcapsules not only indicated the exact location of the microcrack and repaired it, but also delineated the width of the microcrack.

A 2-dimensional discrete element method had been used[178] to investigate the interaction between a microcapsule and a microcrack. A new type of displacement field near to the crack, and a new sort of coalescence crack, were observed. The results indicated that the coalescence of the crack is affected by the hole size. The elastic modulus, compressive strength and coalescence stress decrease with increasing hole radius. It was suggested that the initial damage imposed during experiments should be greater than 95% of the compressive strength in order to enhance the self-healing effect. Large microcapsules require little initial damage. Another approach[179] simplified the irregular cracks, generated in concrete, into linear regular hexagonal cracks in a 2-dimensional plane and planar cracks in 3-dimensional space. A probability model which described the interaction between cracks and capsules was then based upon the probability theory of integral geometry. The results showed that this approach was suitable for optimizing the capsule content which was required in order to repair cracks in concrete. The mechanical properties of the interfaces between self-healing agents and the matrix in encapsulation-based self-healing concrete, markedly affect recovery. Experimental and numerical investigations have been made[180], of healed-crack concrete under uniaxial tension, in order to investigate interface bonding and self-healing agent distribution on the crack surface. The bonding behaviour of the crack interface depends upon the content of healing agent and upon the mechanical properties of the crack surface. Based upon the experimental results, a numerical model for the interface between the self-healing agent and the matrix was developed in order to investigate the effects of aggregate, pore and interface properties upon the bonding behaviour of the crack interface. This model could be used to assess the recovery efficiency of smart self-healing concrete structures. Finite-element analysis was used[181] to simulate the breakage of the capsules. A 2-dimensional circular capsule having various core/shell thickness ratios, embedded in a mortar matrix, was analyzed numerically together with its interfacial transition zone. Zero-thickness elements were incorporated into the solid in order to represent potential cracks. Attention was focused upon the effects of mismatched fracture strength and energy, between capsule and matrix, on the probability of capsule breakage. Extensive simulations of 2-dimensional specimens under uniaxial tension revealed key features of the fracture behaviour, as summarised by fracture maps (figure 16).

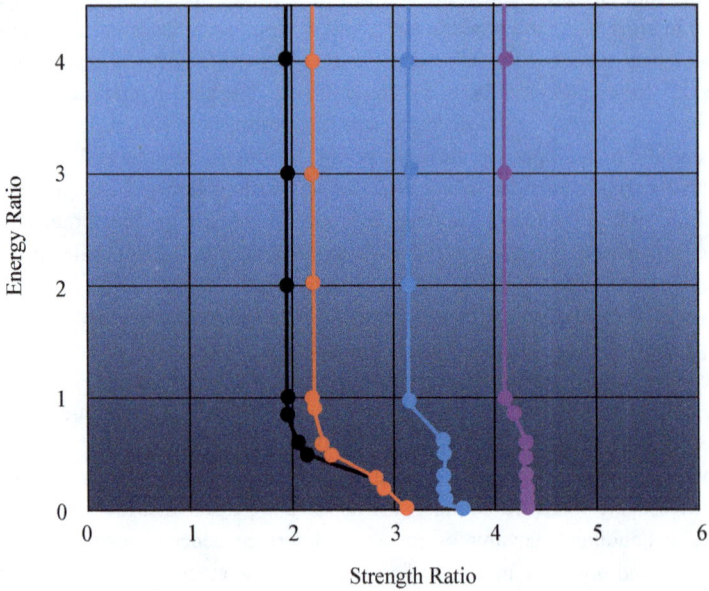

Figure 16. Fracture map for a model microcapsule-containing concrete. Core/shell radius-ratio: black = 1, red = 5, blue = 10, purple = 15. Capsule fracture to left of line, no fracture to right

The cement sheaths which are used in oil and gas production may crack in locations which are difficult to locate and reach, but repair is possible by using self-healing materials[182]. Potassium sodium tartrate exerts a self-healing effect but, when directly added to Portland cement, easily causes marked retardation. Micro-encapsulation of the tartrate in urea-formaldehyde resin microcapsules is therefore indicated. The particle size could appreciably affect particle gradation between microcapsules and cement. Microcapsules which were prepared by optimizing the agitation rate led to acceptable mechanical properties and excellent particle grading. The microcapsules permitted excellent healing of cracks having widths of less than 86μm.

In the search for polymers with which to fill healing capsules, a range of imidazolium-based norbornene polymerized ionic liquids having side-chain structures were synthesized[183]. Inserted imidazolium groups divided the side-chains into two types, termed spacer and tail. By altering the lengths of these two types independently, the self-healing efficiency of the polymerized ionic liquids could be greatly improved without impairing the mechanical strength. Increasing the spacer segments by 8 methylene units decreased the glass transition temperature by 70C and changed the polymerized ionic liquid from a strong material to a very extendable material. Adjusting the lengths of the spacer and tail could increase the mechanical strength. Long-tail segments having 5, 7 or 9 methylene units formed an additional tail region between the ionic aggregates, which then markedly reduced the average aggregation distance and thus accelerated the healing kinetics. It was therefore possible to optimise simultaneously both the mechanical strength and the healing efficiency.

Bacteria

In an early study, the possibility of using bacteria to act as self-healing agents in concrete was explored by using a group of alkali-resistant spore-forming bacteria which was related to the *Bacillus* genus. It was found that spores which were directly added to the cement component (figure 17) remained viable for up to 4 months. A decrease in the pore diameter during setting could limit the lifetime of the spores because it could fall below 1μm; the typical *Bacillus* spore size. The bacteria act, in effect, like catalysts and a suitable mineral compound has to be added in order to provide an autonomous repair mechanism. The maximum amount of mineral which can be added may nevertheless be limited in order to avoid adversely affecting other properties. It was shown[184] that the addition of bacteria plus calcium lactate led to the formation of large amounts of 20 to 80μm precipitates on the crack surfaces of fresh, 7-days cured, specimens. Precipitates are likely to be based upon calcium carbonate and to form due to the metabolic conversion of calcium lactate, according to:

$$CaC_6H_{10}O_6 + 6O_2 \rightarrow CaCO_3 + 5CO_2 + 5H_2O$$

The yield of calcium carbonate-based minerals will increase further when the product CO_2 molecules react with portlandite: $Ca(OH)_2$. That is,

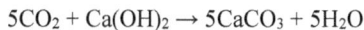

$$5CO_2 + Ca(OH)_2 \rightarrow 5CaCO_3 + 5H_2O$$

This resembles a slow process which occurs in concrete due to the inward diffusion of atmospheric carbon dioxide. The bacterial mineral formation from calcium lactate is an alternative mechanism to an urease-based one, but here the metabolic conversion of calcium lactate does not result in the production of huge amounts of ammonia. The action of the present, bacteria plus lactate, healing agent appeared to be limited to fresh concrete. The copious production of larger precipitates was observed in 7-days cured, but not in 28-days cured specimens. The suitability of chosen bacteria is related to their ability to form spores, which can withstand high mechanical forces and exhibit long-term viability. A solution to the possible loss of viability is to encapsulate or immobilize spores in a protective matrix before adding them to the mix. Among possible solutions is to embed bacterial spores in sol–gel materials.

In further work[185], the self-healing mixture of bacterial spores and calcium lactate was embedded in porous expanded clay particles which acted as reservoirs. Upon crack formation, bacterially-driven calcium carbonate formation (figure 18) led to the closure of micro-cracks. Some 82% of cracks were entirely healed in the case of bacteria-based specimens, while crack-healing fell to 61% in control specimens. Oxygen-consumption measurements indicated that the bacteria were still active up to 9 months after casting. Subsequent bacterially-controlled calcium carbonate formation resulted in the physical closure of micro-cracks[186]. Healing of cracks which were up to 0.46mm wide was observed in bacterial concrete, but only cracks which were up to 0.18mm-wide were healed in control specimens after 100 days of submersion in water. The fact that the improved crack-healing was due to the metabolic activity of bacteria was confirmed by oxygen measurements which revealed the occurrence of O_2 consumption by bacteria-based specimens but not by control specimens.

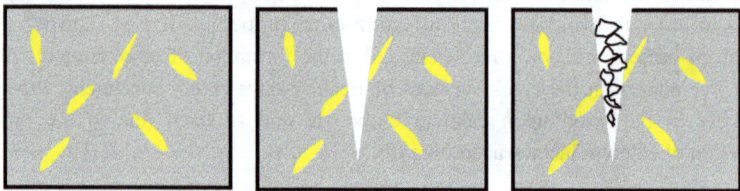

Figure 17. Schematic of steps in the self-healing of biocement
gray: cement, yellow: dormant bacteria, white: CaCO₃ crystals

Because bacterial activity greatly decreases in the high-pH (> 12) environment of concrete, the use of silica gel or polyurethane to protect bacteria was investigated[187]. Experiments showed that gel-immobilized bacteria exhibited more activity than did polyurethane-immobilized bacteria. More (25wt%) $CaCO_3$ therefore precipitated in silica gel than did in polyurethane (11wt%). Cracked mortar specimens which were healed by polyurethane-immobilized bacteria exhibited a greater (60%) restoration of strength and a lower (10^{-10} to 10^{-11}m/s) water permeability coefficient when compared with specimens that were healed by gel-immobilized bacteria. In those samples, the restoration of strength was only 5% and the water permeability coefficient was 10^{-7} to 10^{-9}m/s. Diatomaceous earth has also been used[188] to protect bacteria from the high-pH environment of concrete. Thus-immobilized bacteria exhibited a much higher ureolytic activity, with 12 to 17g/l of urea being decomposed within 3 days, than that of non-immobilized bacteria where less than 1g/l of urea was decomposed within the same time-span. The optimum concentration of diatomaceous earth which was required for immobilization was 60% (w/v) per volume of bacterial suspension. Cracks which had widths ranging from 0.15 to 0.17mm, in specimens containing earth-immobilized bacteria, were completely filled by precipitation. Scanning electron microscopy confirmed that the precipitation around the crack wall was calcium carbonate. Capillary water-absorption tests showed that specimens with earth-immobilized bacteria exhibited the lowest water absorption; 30% of that of control samples. Precipitation within the cracks thus increased the water penetration resistance of cracked specimens.

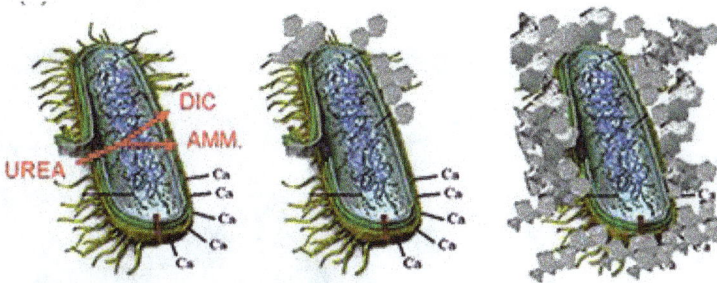

Figure 18. Mechanism of biological self-healing. Reproduced (part of larger array) from Status-of-the-Art for Self-Healing Concrete, Li, X., Zhao, S., Wang, S., Journal of Physics - Conference Series, 1622, 2020, 012011 under Creative Commons Licence 3.0.

In a typical study[189], the degree of self-healing is judged by the crack width. This showed that, following curing for 40 days, cracks in concrete could be filled with calcium carbonate that was produced by microbial mineralization. The self-healing effect was evident, and the maximum width was greater than 1mm. The amount of calcium carbonate was greatest at the opening of the crack. This quantity gradually decreased with increasing crack depth. When the depth was greater than 10mm, calcium carbonate was no longer found. When the distance was less than 1.5mm, micro-organisms could produce a large amount of calcium carbonate via mineralization but, when the distance was greater than 1.5mm, the calcium carbonate gradually decreased with increasing depth because the microbial mineralization process required oxygen. The general method is to incorporate bacteria into the concrete matrix in order to heal cracks as soon as they appear. The self-healing capability depends upon the type of bacteria, the substrate type and the crack size. Also relevant is the number of living bacteria within the concrete, the curing condition and the curing time. Microbially generated calcite precipitation was again investigated[190] via control of the pH level, the oxygen content and the substrate. The results showed that the pH value of the bacterial liquid gradually increased from 7.0 to 8.3 as the bacteria grew. Calcite formed due to bacterial metabolic conversion rather than due to the direct decomposition of the substrate under the influence of extracellular enzymes. The metabolic process produced CO_2, the substrate provided Ca^{2+} and CO_3^{2-} and other organic nutrients also provided CO_3^{2-}.

Table 6. Compressive Strength of Cubes

Rice Husk (%)	Bacteria (l/m³)	Curing (days)	Compressive Strength (N/mm²)
0	0	7	34.55
0	0	28	44.2
5	0	7	9.45
5	0	28	18.1
0	100	7	38
0	100	28	49.55
5	100	7	11.7
5	100	28	22.7

It was found that the growth of bacteria created an alkaline environment and produced CO_2. The cell walls of bacteria were negatively charged and affected calcium carbonate precipitation by acting as sites for nucleation or calcium enrichment. The CO_2 molecules which were produced reacted with Ca^{2+} to form calcite crystals in an alkaline environment. Further investigation[191] showed that bacteria-related $CaCO_3$ precipitation could lead to crack-plugging and to a decrease in the permeability of cement-based materials. There were two main types of self-healing, depending upon the bacterial species. In one type of self-healing, based upon spore-forming alkali-resistant bacteria, spores would be active when cracks formed at the concrete surface and would then decompose calcium lactate so as to form calcium carbonate. The other type of self-healing was based upon producing urease bacteria so that bacteria which were incorporated into the concrete decomposed urea so as to form carbonate ions which - in the presence of calcium ions – led to $CaCO_3$ being precipitated in the cracks.

A mathematical model for the bacterial self-healing of a crack was used to investigate the ability of bacteria to catalyze the process in concrete[192]. Spherical clay capsules which contained calcium lactate and nutrients were assumed to be embedded in a concrete structure. Water which entered a new crack released the capsule contents and caused the bacteria to convert calcium lactate into calcium carbonate. The crack was then sealed due to metabolically driven limestone precipitation. The model was formulated as a moving-boundary problem in which two fragments of boundary moved due to calcium carbonate precipitation and to dissolution of the capsule contents.

A study was made[193] of indigenous micro-organisms, *Proteus mirabilis* and *Proteus vulgaris*, which can be isolated from soil. They were able to precipitate calcium carbonate, and broken concrete was treated with a culture which contained the micro-organisms. It was found that the cracked concrete could thus be filled with calcium carbonate, but the treatment had little effect upon the strength of the concrete.

Table 7. Split tensile strength of cylinders

Rice Husk (%)	Bacteria (l/m³)	Tensile Strength (N/mm²)
0	0	2.99
5	0	2.41
0	100	3.34
5	100	2.73

A study of the production of spores of alkaliphilic *Bacillus cohnii* showed[194] that the optimum carbon source was sucrose at a concentration of 1.0g/l, and that the optimum nitrogen source was beef extract with a concentration of 3.0g/l. The optimum concentration of Mn^{2+} was 3.2mg/l and the optimum concentration of Mg^{2+} was 0.12g/l.

A suitable preparation temperature was 30C. The carbon source, nitrogen source and the Mn^{2+} content were the most important factors. The maximum expected spore production rate was 1.67 x 10^9/ml when the sucrose, beef-extract and Mn^{2+} concentrations were 1.30g/l, 3.29g/l and 11.48mg/l, respectively. The actual spore production rate was 1.50 x 10^9/ml. The feasibility of using expanded perlite for the immobilization of *Bacillus cohnii* was demonstrated[195] with respect to the direct introduction of bacteria and to the use of expanded clay to immobilize bacteria. Experimental data showed that specimens which contained perlite-immobilized bacteria exhibited the most efficient crack-healing behaviour within a given period. The crack widths which were completely healed after 28 days ranged up to 0.79mm, as compared with the value of 0.45mm for specimens which contained clay-immobilized bacteria. Precipitates on the crack surfaces were found to be calcite crystals. Coated expanded perlite was used[196] to immobilise bacterial spores and encapsulate nutrients as two separate components for self-healing. Healing could be achieved when coated expanded perlite containing self-healing agents was used as a 20% replacement of fine aggregate and if a suitable ratio of spores to calcium acetate was present. This showed that self-healing is not simply a question of providing sufficient healing compounds, such as calcium acetate, but that a minimum number of bacterial spores is also required to ensure that sufficient cells participate in the healing process. The effects of various carbon sources, nitrogen sources, Mn^{2+} concentrations, temperature, pH and nutrients upon the germination of spores of the alkaliphilic *Bacillus cohnii* were more recently studied[197]. The total cell-count and sporulation rate of the bacteria were 3.14 x 10^9cfu/ml and 92.6%, respectively, under optimum conditions. The optimum pH value for spore germination was 9.7. While Na^+ strongly stimulated germination of the spores, other cations such as K^+, NH^{4+} and Ca^{2+} did not. The optimum Na^+ concentration was 200mM. The germination rate of spores in a control specimen at 24C was more than 50%, and the spores had a mineralizing effect following germination, with the crystals produced by microbial-induced precipitation being pure calcite. When the crack width in specimens containing the spores was less than 1.2mm, it could be completely repaired by 28 days of healing.

Table 8. Flexural Strength Test of Prisms

Rice Husk (%)	Bacteria (l/m³)	Flexural Strength (N/mm²)
0	0	6.62
5	0	5.64
0	100	6.87
5	100	5.88

In another study[198] of the use of *Bacillus cohnii* to precipitate calcite, fine aggregate was partially replaced by rice husk. Experiments were carried out using compressive, split tensile, flexural and water-absorption tests. The compressive strength after 7 days was increased by 10% (table 6) due to adding *Bacillus cohnii*. The 28 day compressive strength was increased by 12%. The compressive strength after 7 days, if the fine aggregate was replaced by 5% of rice husk was increased by 23%, while the 28 day compressive strength was increased by 26%. The bacterial activity was greater when rice husk was added. The split tensile strength after 28 days was increased by 12% (table 7) by adding *Bacillus cohnii* and by 14% when the fine aggregate was partially replaced by 5% of rice husk. The flexural strength after 28 days was increased by 4% (table 8) and, when the fine aggregate was partially replaced with 5% of rice husk, it was increased by 7%. Water-absorption tests (table 9) showed that the absorption rate was greater in samples which contained rice husk, but the addition of bacteria led to a reduction. A further study[199] of the use of high-porosity expanded perlite as a bacteria carrier showed that a pure culture of *Bacillus cohnii* and a microbial mixture of aerobic, facultative anaerobic and anaerobic bacteria all led to excellent crack-healing. The healing-rate of concrete after curing for 28 days was 73.3% for *Bacillus cohnii*, 83.3% for aerobic bacteria, 63.3% for facultative anaerobic bacteria and 41.5% for anaerobic bacteria. The final precipitates were all forms of calcium carbonate crystals.

A sugar-coating method for immobilizing bacteria and nutrients was proposed, and the effect upon crack-healing of expanded perlite which was wrapped with various materials as a bacteria and nutrient carrier was investigated[200]. Wrapping materials could effectively reduce the breakage rate of the perlite particles and resist the intrusion of water when mixing with concrete. Following wrapping the perlite particles increased the healing of concrete cracks. The crack-healing in fact peaked when perlite particles which were immobilized with bacterial spores were wrapped with low-alkali potassium magnesium phosphate cement. This was confirmed by water permeability experiments. After 28 days of crack-healing, the maximum width of a completely healed crack was 1.24mm.

A protective carrier for bacteria was based[201] upon calcium sulpho-aluminate cement, a low-alkali rapid-hardening cementitious material. By adjusting the carrier composition and the healing-agent content, the compatibility of the carrier with both the healing agent and the concrete could be optimized. The carrier preserved bacterial activity over long periods of time. When this self-healing system was embedded in concrete, cracks which were up to 417μm wide exhibited nearly 100% crack closure within 28 days. As compared with plain mortar, the restoration ratios of the compressive strength and water-tightness were 130 and 50%, respectively. When calcium sulpho-aluminate cement with

20% silica fume was used, self-healing was successful; with maximum crack widths of about 322μm being healed within 28 days[202]. Following self-healing, the concrete compressive strength had recovered up to 84% of its original value and the water-tightness had been completely restored. Again using low-alkali sulfo-aluminate cement as a carrier to encapsulate spores, the effects of the spore group and microbial group upon the behaviour of the concrete were studied[203]. The area repair ratio, the water permeability, the repair ratio of anti-chloride ion penetration and the ultrasound velocity were used to evaluate the self-healing efficiency of cracks after various curing ages. The growth, enzyme activity and microbial morphologies of spores, with and without encapsulation, were used to evaluate the protective effect of the carrier with regard to spores. This showed that the addition of the two (spore and microbe) self-healing agents could slightly affect the basic concrete behaviour. In early-stage cracks, the two types of self-healing agent could impart a good self-healing effect. In the case of later-stage cracks, samples in the microbe group could still be repaired well, while self-healing of samples in the spore group was greatly reduced.

Table 9. Water absorption

Rice Husk (%)	Bacteria (l/m^3)	Absorption (%)
0	0	4.58
5	0	10.1
0	100	3.4
5	100	7.04

Microbial spores of *Bacillus megaterium*, nutrients and low-alkaline carriers were combined[204] by extrusion and low-alkali calcium sulpho-aluminate cement was used as a carrier to protect the spores from the high alkalinity which was caused by cement hydration. The integrated self-healing agent was incorporated into concrete, and the crack-healing efficiency was judged on the basis of area repair-ratio, water-resistance recovery, the recovery of chloride-ion resistance and the healing depth. Two different nutrients were used as precursors for microbial mineralization, showing that *Bacillus megaterium* could use glucose and calcium lactate as precursors, and that this new repair system markedly improved the self-healing efficiency. Calcium lactate had a better repair effect than that of glucose, when used as a precursor, and the average healing depth reached 4000μm. The calcium sulpho-aluminate cement was an excellent spore carrier and could attenuate the negative effect of nutrients on concrete when micro-organisms and nutrients were assembled with this sulpho-aluminate cement.

The feasibility of using rubber particles as a bacteria carrier in self-healing concrete was examined[205]. Two types of self-healing concrete were prepared, having rubber particles of differing sizes. The mechanical properties of the self-healing rubber concrete were compared with those of plain concrete and normal rubber concrete. This showed that self-healing rubber concrete with a particle size of 1 to 3mm possessed a better healing capacity than that of self-healing rubber concrete with a particle size of 0.2 to 0.4mm, with the maximum width of a completely healable crack being 0.86mm. The compressive strengths of the self-healing rubber concrete were low to begin with, but then exceeded those of normal rubber concrete at 28 days. The self-healing rubber concrete had a higher splitting tensile strength than did plain concrete, and possessed a better anti-crack capability.

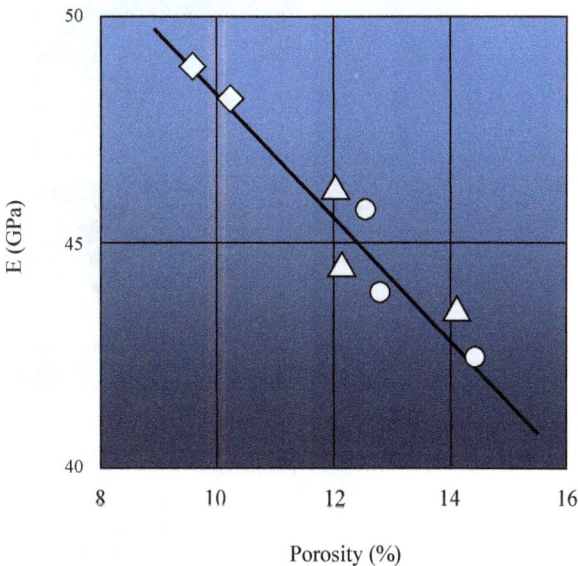

Figure 19. Young's modulus as a function of concrete porosity and bacterial/nutrient content. Circles: control, triangles: nutrient, diamonds: bacteria

The use of biochar as a carrier for carbonate-precipitating bacteria spores was evaluated[206]. Super-absorbent polymer, and polypropylene microfibers, were added in order to provide moisture to the bacteria and to control crack-propagation. Samples were damaged by pre-loading to 50 or 70% of the peak strength. The biochar-immobilized spores, combined with super-absorbent polymer and polypropylene, precipitated sufficient amounts of calcium carbonate to seal completely cracks which were up to 700μm wide. This combination also led to the greatest restorations of impermeability and strength for both levels of pre-loading. An improvement in strength of 38% and reductions in water penetration and absorption of 65 and 70%, were found for the immobilization of spores in biochar; as compared with directly-added spores.

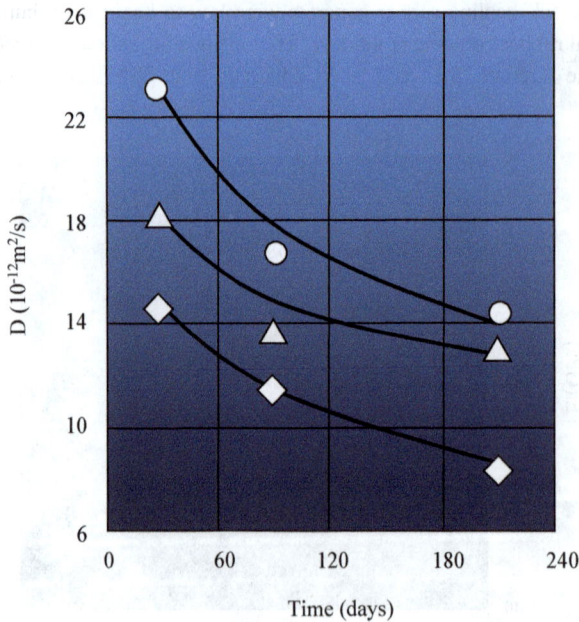

Figure 20. Non-steady chloride migration coefficient as a function of time. Circles: control, triangles: nutrient, diamonds: bacteria

The inclusion of polypropylene fibers contributed to the restoration of strength and impermeability, while the super-absorbent polymer ensured a higher volume of bacteria-induced carbonate precipitation. The use of biochar could also be considered to be a carbon-sequestration method and therefore advantageous for the control of global warming. A further comparison was made[207] of autogenous and other self-healing methods by using steel plus polyvinyl acetate fibers, super-absorbent polymer or bacteria which were immobilized in biochar. Swelling of the super-absorbent polymer upon exposure to water, and hydration enhancement by the polymer, led to an improved blocking and filling of cracks. The effectiveness of crack closure by the polymer alone or by the polymer plus fibers, was limited to surface cracks that were less than 500μm wide. Microbial calcite precipitation by biochar-immobilized bacteria, combined with super-absorbent polymer and fiber, gave better closure of surface cracks which were greater than 600μm wide and of internal micro-cracks as compared with the other methods. The effectiveness of crack-filling by immobilized spores in biochar was greater than the direct addition of spores or super-absorbent polymer. The precipitation of calcium carbonate crystals in internal cracks and interfacial zones around polyvinyl acetate fibers, in concrete with biochar-immobilized spores, led to a good restoration of strength and permeability as compared with autogenous healing. Macro-voids which were formed by super-absorbent polymer with a larger average particle size and higher swelling capacity affected the total permeability and permeability recovery following repeated healing.

Bacillus sphaericus spores, encapsulated in an alginate-based hydrogel, remained viable after encapsulation and could precipitate a large amount, some 70wt%, of $CaCO_3$ in the hydrogel matrix[208]. The encapsulated spores were added to mortar and the *in situ* bacterial activity was monitored by determining the oxygen consumption of a simulated crack surface. Specimens with added free spores exhibited no oxygen consumption, thus confirming protection of spores by the hydrogel. The best technique for encapsulating spores of *Bacillus sphaericus* in sodium alginate in order to protect them during concrete mixing was explored[209]. Extrusion, spray-drying and freeze-drying were used to encapsulate the spores in the alginate. Freeze-drying led to the highest (100%) spore survival rate while extrusion and spray-drying led to survival rates of 93.8 and 79.9%, respectively. The results also revealed that freeze-dried spores exhibited the highest level of urea decomposition. The crack-healing ratio for mortar with freeze-dried micro-encapsulated spores was significantly higher than the ratio for samples without bacteria.

It was shown that alkaliphilic *Bacillus subtilis* is able to survive for long periods, in concrete or mortar, due to its spore-forming ability. *Bacillus subtilis* bacteria were introduced into concrete by direct incorporation or via various carriers such as lightweight aggregate and graphite nano-platelets[210]. In each case, calcium lactate was used as an organic precursor. Bacteria which were immobilized in graphite nano-platelets gave better results for specimens which were pre-cracked at 3 and 7 days. Bacteria which were immobilized in lightweight aggregates were more effective in samples that were pre-cracked at 14 and 28 days.

Table 10. Compressive strength of concrete with bacteria and polypropylene fibers

Sample	Curing (days)	Compressive Strength (N/mm²)
plain	7	16.5
plain	14	23.37
plain	28	27.22
bacterial	7	17.63
bacterial	14	25.42
bacterial	28	21.29
fiber-reinforced	7	19.15
fiber-reinforced	14	27.98
fiber-reinforced	28	32.35
bacteria plus fibers	7	22.96
bacteria plus fibers	14	31.85
bacteria plus fibers	28	37.87

The bacteria which were immobilized in lightweight aggregate also imparted a large increase in the compressive strength to concrete. Although specimens which had lightweight aggregate as the carrier were not as efficient as graphite in the early stages, they were consistent in their crack-healing efficiency when the samples were pre-cracked later. Specimens which were directly incorporated with bacteria did not exhibit any crack-healing effects. It was concluded that the addition of *Bacillus subtilis* resulted in a slight increase in compressive strength, regardless of the incorporation technique.

Genetic engineering was used to transfer the bioremediase-like gene of a thermophilic anaerobic bacterium to the *Bacillus* strain in order to create a true self-healing biological agent. Incorporation of the transformed bacterial cells into concrete or mortar led to higher mechanical strengths and to an improved durability, as compared with those of normal concretes[211]. Microstructural analysis revealed the formation of a new gehlenite phase, $Ca_2Al_2SiO_7$, as well as the deposition of calcite within transformed *Bacillus subtilis* treated samples. Gradual development of a nano-rod gehlenite composite within the specimens was attributed to the biochemical activity of bioremediase-like protein within the incorporated cells. When peptone, yeast extract and *Bacillus subtilis* were added to a concrete mix[212], there was a decrease in porosity and an increase in strength and dynamic modulus (figure 19) together with a reduction in water uptake, gas permeability and chloride permeation. The microbial precipitates in cracks were $CaCO_3$, and the morphologies of these calcite crystals ranged from needle-like to bouquet-like and rhombohedral. A crack with a surface width of 400μm was completely filled after 44 days. The chloride ion diffusivity was much reduced by adding the bacterium and the yeast-extract plus peptone medium (figure 20).

Table 11. Flexural strength of concrete with bacteria and polypropylene fibers

Sample	Curing (days)	Flexural Strength (N/mm²)
plain	7	7.1
plain	14	7.67
plain	28	8.34
bacterial	7	7.4
bacterial	14	8.1
bacterial	28	8.55
fiber-reinforced	7	7.8
fiber-reinforced	14	8.2
fiber-reinforced	28	8.7
bacteria plus fibers	7	8.05
bacteria plus fibers	14	8.63
bacteria plus fibers	28	9.12

The effect of the microbial additions lasted up to 28 days, and no appreciable changes were observed up to 210 days. The changes in pH showed that the bacterial concrete had not absorbed a lot of Cl⁻ ions, due to the lower porosity and gas permeability in comparison with that of control samples. The transmission of chloride was used[213] to evaluate the protective effect of the microbial self-healing of cracks. Electromigration was used to accelerate the chloride transmission. The corrosion current and self-corrosion potential data showed that the degree of corrosion of reinforcement was lower when the cracks were blocked by self-healing, while fewer corrosion products overflowed from the cracks. The main reason was that the healed cracks hindered the chloride from penetrating the specimens via the cracks and slowed reinforcement corrosion. The chloride contents at various depths showed that the chloride content in self-healed specimens was lower when the cracks were healed. It was confirmed that microbial self-healing can impede chloride transmission in cracks, thus protecting the reinforcement.

Table 12. Split tensile strength of concrete with bacteria and polypropylene fibers

Sample	Curing (days)	Tensile Strength (N/mm²)
plain	7	1.63
plain	14	2.05
plain	28	2.53
bacterial	7	1.7
bacterial	14	2.28
bacterial	28	2.72
fiber-reinforced	7	1.82
fiber-reinforced	14	2.37
fiber-reinforced	28	2.85
bacteria plus fibers	7	1.93
bacteria plus fibers	14	2.58
bacteria plus fibers	28	3.19

The characteristics of *Bacillus subtilis* which are relevant to the self-healing of concrete were modified genetically[214] by incorporating various genes: **gerA** encodes germination receptors which are activated by L-alanine, while **tupA** is responsible for the synthesis of teichuronopeptide and **ca** encodes the carbonic anhydrase which catalyzes the synthesis of carbonate ions. In order to protect the bacterial cells from being squeezed, microspheres were prepared using microcrystal cellulose before introducing the bacteria. It was found that samples which expressed **gerA** exhibited a 39.9% germination ratio, as compared with the 17% of the host cells. The insertion of **tupA** imparted a higher resistance to alkaline environments and the samples tolerated pH values as high as 11, as compared with the original strain and its limiting pH of 9. The **ca**-modified strain promoted more calcium carbonate production than did the original material. *Bacillus sphaericus* and *Bacillus subtilis* were incorporated into concrete which was reinforced with polypropylene fibres, and artificial cracks were created[215].

The results showed that the combination of bacteria and fibres increased the compressive strength by 39.12% (table 10), the flexural strength by 9.35% (table 11) and the split tensile strength by 26.09% (table 12). Some 70 to 80% of the strength was restored within 28 days. There was a good bonding between the fibres and the bacteria. Cracks self-healed completely within 28 days. There was a deposit of calcium carbonate over the cracks, and quartz was also present. Direct introduction and immobilization were used[216] to incorporate *Bacillus subtilis*. In the case of immobilization, iron oxide nano/micro particles or bentonite nano/micro particles were employed. The effectiveness was monitored by performing compression and tensile tests after 3, 7 and 28 days. Immobilization using iron oxide nano/micro particles was best, and led to the healing of cracks which were up to 1.2mm wide, together with an 85% recovery of the compressive strength. The use of bentonite nano/micro particles, or direct introduction, permitted the healing of cracks with widths of only 0.15 and 0.45mm, respectively, with corresponding restorations of strength of 45 and 65%. *Bacillus subtilis* was added[217] to engineered cementitious composites, normal mortar and concrete, together with calcium lactate, urea and yeast-extract as a nutrient. The mixture was first immobilized in pumice solution or porous expanded clay. Bacterial cell solutions with concentrations of 10^4, 10^6 or 10^8cells/ml were mixed with mortar and concrete in ratios of 1:1:2 or 1:1.32:2.5, with a water/cement ratio of 0.5. These cell solutions were also mixed with engineered cementitious composites with a water/binder ratio of 0.33 and a sand/binder ratio of 1:0.84. As usual, the bacteria reacted with the concrete components to form calcium carbonate precipitates or limestone which sealed the cracks. From a bacterial concentration of 10^4cells/ml, the compressive strength of concrete increased to a maximum at 10^6cells/ml and had declined at 10^8cells/ml. The use of stiffen-sand encapsulated, pumice-immobilized and loaded clay bacteria had increased the compressive strength of concrete by 35.20, 31.59 and 0.20%, respectively, after 28 curing days. Cracked stiffen-sand capsules and pumice-immobilized engineered cementitious composite specimens regained flexural strengths of 23.6 and 32.22% after 56 self-healing days. Bacteria-incorporated concrete specimens with crack widths of 0.015, 0.018 and 0.02mm were sealed by $CaCO_3$ precipitates after 56 self-healing days.

Materials Research Forum LLC
https://doi.org/10.21741/9781644901373

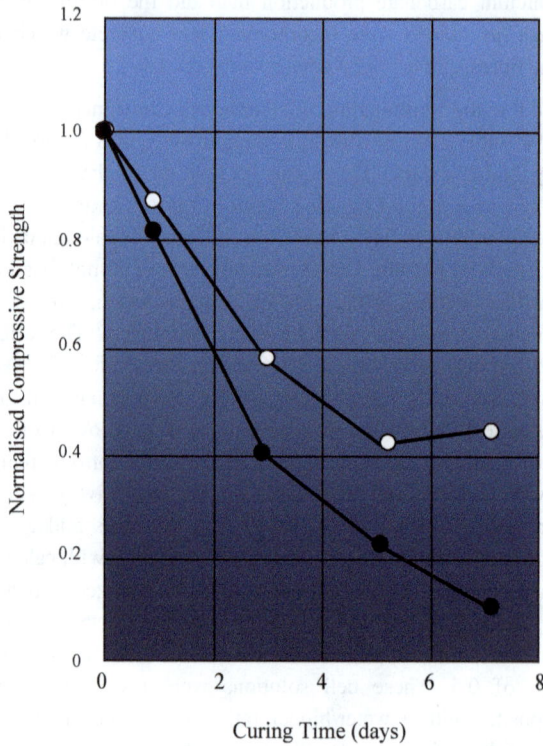

*Figure 21. Normalised compressive strength of concrete
cured in pure water (black) or bacterial solution (white)*

Another comparison[218] of immobilizing media involved iron oxide nano-particles, limestone micro-particles and milli-sized siliceous sand. Among these, the iron oxide particles were found to be the most effective immobilizer for protecting *Bacillus subtilis*, followed by the siliceous sand and limestone. The magnetic immobilization of bacteria with iron oxide nanoparticles, and the effect upon the compressive strength and drying shrinkage was specifically investigated[219]. The results indicated that the addition of immobilized *Bacillus* species with iron oxide nanoparticles contributed to an increase in compressive strength. The precipitated crystals which were observed were those of $CaCO_3$. The effect of magnetic immobilized cells upon concrete water absorption showed that concrete containing bacteria with iron oxide nanoparticles exhibited a higher resistance to water penetration[220]. The initial and secondary water absorption rates in the bio-concrete were 26 and 22% lower than those of control specimens. As another refinement, 3-aminopropyltriethoxy silane-coated iron oxide nanoparticles were used[221] as a biocompatible carrier for *Bacillus* species. The presence of 100μg/ml of the new particles could increase bacterial viability, and the greatest $CaCO_3$ production was observed when the cells were decorated with 50μg/ml of the particles.

In a further study[222], bacteria were immobilized using recycled coarse aggregate and virgin fine aggregate. Vegetative cells of *Bacillus subtilis* were incorporated into recycled coarse aggregate by vacuum impregnation, and the crack-healing efficiency was studied by measuring healed-crack widths and the percentage restoration of strength after 3, 7 and 28 days. Specimens with recycled coarse aggregate and 50% virgin fine aggregate exhibited the healing of cracks with widths of up to 1.1mm, with 85% recovery of the compressive strength. Specimens with only recycled coarse aggregate exhibited a maximum value of 0.7mm for healed cracks, and a 76% recovery of strength. The direct, non-immobilized, incorporation of bacteria yielded values of a 0.6mm width for healed cracks and a 69% recovery of strength. The use of recycled coarse aggregate alone decreased the compressive strength by up to 3%. Alkali-treated micro-cellulose fiber has lately been investigated[223] as a *Bacillus subtilis* carrier for self-healing mortar. Two types have been examined. In one type, nutrients were added to the mortar mix while, in the other, the nutrients were added to the curing water. Self-healing mortar with cellulose fiber as the bacteria-carrier resulted in the greatest self-healing, as compared with other compositions: 8.23% more than that of pre-cracked control samples after 28 days. The fiber increased the availability of bacteria in a cracked region by acting as a bridge across the crack. At the dosages used, the addition of cellulose fiber led to a decrease in the compressive strength. A 6.97% decrease in the compressive strength was noted upon adding 0.5% fibers. The decrease was attributed to a loss in workability of the mortar, which led to lower compaction of the samples. A greater decrease in compressive strength was observed for compositions which contained calcium lactate. This was because the lactate did not directly take part in the hydration of cement. The by-product, calcium carbonate, instead caused a reduction in compressive strength. Samples which contained fibers and bacteria, and which were cured in calcium lactate, exhibited a lesser decrease in compressive strength. The loss in strength occurred only for mixes with 0.5% of fiber and with 1.3×10^7 cells/cm^3. The fibers could also act as water reservoirs for internal curing. In order to judge the effect of soil incubation upon self-healing process, experiments were performed[224] on mortar which was impregnated with *Bacillus subtilis* that was encapsulated in calcium alginate. The specimens were cracked, and sub-divided into three groups. Each group was incubated for 28 days in partially-saturated soil, fully-saturated soil or water. This showed that self-healing could be activated, in cracks which were exposed to saturated soil, provided that the matrix suction was lower than the capillary pressure of the crack. There was moreover no evidence that naturally existing bacteria in the soil had an effect upon the self-healing process.

The potential use of natural fibers, such as coir, flax and jute, to carry the bacterial spores has been explored[225]. Calcite-precipitating bacteria such as *Bacillus subtilis*, *Bacillus cohnii* and *Bacillus sphaericus* were introduced into the concrete together with calcium lactate pentahydrate and urea as nutrients. The results suggested that natural fibers were capable of substantially immobilizing bacterial spores. The flax fibers better protected the bacteria, with improved crack-healing and recovery of compressive strength, while the coir fibers led to a higher compressive strength. The *Bacillus sphaericus* precipitated more calcium carbonate, with uniform healing over the entire length of cracks. The overall healing rates which were imparted by fiber-immobilized bacteria, after 7 and 28 days, were 75 to 85% and 60 to 65% respectively for pre-cracked specimens. The mechanism involved in the coupled effect of fibers and bacteria has been investigated[226], confirming that the coupled effect could result in a higher repair efficiency, better mechanical properties and the recovery of water resistance. Biological staining techniques were used to observe the role played by biological organic matter in the repair process.

Table 13. Compressive strength of concrete with various contents of bacteria-containing beads

Beads (%)	Compressive Strength (MPa)
0	37.31
2	35.50
5	33.33
10	30.06
15	22.19

The effect of calcium lactate upon the compressive strength and self-healing ability of bacterial concrete was investigated[227] because the production of calcium carbonate by bacterial action is limited to the calcium content of the cement, and calcium lactate can serve as an additional source of calcium. *Bacillus subtilis* (2,000,000cfu/g with 0.5% cement) was added to concrete in both spore and cultured form. Calcium lactate in concentrations of 0.5, 1.0, 1.5, 2.0 or 2.5% of cement was also added. In other cases, cultured *bacillus subtilis* in a concentration of 100,000cells/ml was mixed with the concrete. Cubes (100 x 100 x 100mm) were tested following curing for 7, 14 or 28d. There was a significant increase in the compressive strength of specimens with calcium lactate and bacteria, with a maximum increase of 12% when 0.5% of calcium lactate and bacteria were added. Increasing the percentage of calcium lactate further in fact led to a reduction in the compressive strength, due to such materials not taking part in hydration of the binder. The bacteria and lactate were not directly involved in hydration, but the calcium carbonate by-product helped to fill pores and heal cracks. Multivariate analysis showed that,

$$\text{compressive strength (MPa)} = 19.252 - 3.252x + 2.0212y - 0.045922xy - 0.038322ey^2$$

where x is the percentage of calcium lactate and y is the age (days) of the specimen. Scanning electron microscopy revealed the formation of calcium silicate hydrate, calcite and needle-like ettringite. Following 28 days of curing, bacterial concrete with 0.5% of calcium lactate had a dense compact structure due to $CaCO_3$ production by bacteria, and this resulted in the increase in compressive strength over that of control samples.

The self-healing of concrete using *Bacillus subtilis* natto was re-visited by studying experimentally[228] the effect of biomineralization with lightweight aggregate as a protective vehicle that could control the release of healing fluid during 4 cracking-healing cycles. The bacteria were mixed with a solution of urea and calcium chloride. The results indicated the existence of bacterial $CaCO_3$ which formed following 4 healing cycles. Restoration of the compressive strength (figure 21) confirmed the high self-healing capability of the scheme. A crack-healing method which was based upon biomineralisation was used to repair microcracks in concrete specimens[229]. This repair method proved to be effective and the results showed that the permeability of specimens which were repaired in this way markedly decreased. Samples having a crack width of 0.05mm exhibited a different decreasing trend to that exhibited by samples having a crack width of 0.15mm. The unrepaired crack depths of samples having a crack width of 0.15mm were almost zero and, the smaller the crack width the larger was the unrepaired crack depth. Samples having wider cracks (0.10 and 0.15mm) were repaired more effectively and had higher recovered strengths. This healing method could repair cracks of between 0.05 and 0.15mm while wider (0.10 or 0.15mm) cracks were repaired more completely and with greater effectiveness.

Various *Bacillus* strains, encapsulated in sodium alginate beads, were added to concrete as self-healing agents[230]. The alginate prevented leaching, and the additions – which amounted to between 2 and 3% of the total volume of concrete cubes – could exert their effect without lowering the compressive strength. Calcium carbonate precipitation occurred in the presence of urease activity, and there was a direct involvement of ureolytic bacteria in $CaCO_3$ precipitation. A 16% increase in compressive strength, and appreciable healing, resulted from adding *Bacillus subtilis*. On the other hand, *Bacillus anthracis* produced only a 6% increase in strength after 28 days (table 13). *Bacillus pasteurii* had no appreciable effect, while *Bacillus subtilis* gave better results than any of the other strains (table 14). Crack self-healing was visible to the naked eye, 20 to 30 days after its appearance at the concrete surface. The self-healing involved microbially-induced carbonate precipitation. The bacterial strains were carbonate-producing in the presence of water and generated a calcium carbonate gel, which helped to heal the cracks.

Table 14 Compressive strength of concrete with various bacterial strains

Strain	Curing Time (days)	Compressive Strength (MPa)
control	7	22.54
control	14	32.30
control	28	34.75
Bacillus specie	7	19.85
Bacillus specie	14	28.10
Bacillus specie	28	30.55
Bacillus pasturii	7	21.68
Bacillus pasturii	14	30.47
Bacillus pasturii	28	33.61
Bacillus anthracis	7	22.46
Bacillus anthracis	14	31.14
Bacillus anthracis	28	35.38
Bacillus subtilis	7	25.48
Bacillus subtilis	14	35.32
Bacillus subtilis	28	40.13

Test samples in the form of 150mm x 150mm x 600mm beams were subjected[231] to static loading at their centers, before and after *Bacillus subtillis* injection. Flexural tests were carried out on beams which had been aged for 7, 14, 21 or 28 days. The results showed that *Bacillus subtillis* could be confidently used to repair concrete, given the good flexural strength observed following repair.

The healing ability of the bacterium, *Deinococcus radiodurans*, which is an extremophilic non spore-forming entity has been compared[232] with that of spore-forming *Bacillus subtilis* in mortar using concentrations of 10^3, 10^5 or 10^7cells/ml. Their effect upon the compressive strength and upon water absorption was determined under conditions of both high and low water availability. This showed that *Deinococcus radiodurans* is a potential source of biomineralization as its use could lead to a considerable improvement in the mechanical properties of microbial concrete, equal to that imparted by *Bacillus subtilis* in a favourable environment. In an unfavourable environment, *Deinococcus radiodurans* imparted much better mechanical properties than did *Bacillus subtilis*. The effect of the former bacterium was further tested[233] at 27C and at 4C, using the same concentrations. The results again indicated that its inclusion led to appreciable calcium carbonate precipitation and hence to effective crack-healing at both temperatures. There were marked improvements in compressive strength and water absorption.

The bioprecipitation of $CaCO_3$ crystals by *Bacillus subtilis* has been studied[234] in a semi-solid system, showing that the crystals were produced by the third day of incubation. The main polymorph was calcite, plus some vaterite. Some amorphous $CaCO_3$ was also present. The crystallinity index was 81.1%. This biogenic calcium carbonate did not decrease the pH value and did not lead to chloride formation; in fact, it increased the pH to 10, an advantage when reinforcing concrete.

The spore-forming ureolytic strain, *Bacillus sphaericus*, was again evaluated[235] with regard to its tolerance of alkalies and calcium, its oxygen dependence and its behaviour at low temperatures. Experimental results showed that it had a good tolerance and could grow and germinate within a wide range of alkaline-pH environments. The optimum pH range was 7 to 9. Highly alkaline (pH = 10 to 11) environments slowed, but did not stop, growth and germination. The presence of oxygen was required during bacterial growth and germination, but was not essential during bacterial urea decomposition. The strain also exhibited a good tolerance to calcium; especially at the high bacterial concentration of 10^8cells/ml. No appreciable effect of the presence of 0.9M Ca^{2+}, upon bacterial ureolytic activity, was observed. At 10C, the bacterial spores germinated and revived ureolytic activity with some slight retardation, but that could be countered by using a higher bacterial concentration and by adding yeast extract. It was concluded that this

Materials Research Forum LLC

https://doi.org/10.21741/9781644901373

bacterial strain was a suitable agent for the bacteria-based self-healing of concrete. The conditions obtaining in cement-based materials include high temperatures, a lack of nutrients and a high pH level. These factors can impede microbial-induced calcium carbonate precipitation. Their effect upon the ureolytic bacterium, *Sporosarcina pasteurii*, was therefore studied with regard to bacterial viability and urea hydrolysis[236]. The viability and hydrolysis were most affected by a temperature of 55C and a pH level of 13.6. The impact was much less at 45C and a pH of 12.9. The surface charge of *Sporosarcina pasteurii* cells was slightly affected by all of these test conditions, and its existence suggested that the cells could act as heterogeneous nucleation sites for the calcium carbonate precipitation. Much higher calcite concentrations were indeed observed in bacterial pastes which contained viable cells, as compared with cell-free pastes. Self-healing concrete containing *Bacillus pasteurii* was produced[237] by adding spores and nutrients, with the bacterial content ranging from 10 to 25%, at the time of pouring. The optimum composition led to increases in compressive and flexural strength of 79.16 and 50% respectively; together with a 24.38% increase in sulfate resistance. The highest percentage of calcium carbonate precipitation was equal to 9.49% of the weight of the mixtures, and the associated highest areal repair-rate was able to fill cracks having widths of 0.80mm.

Bacteria-based beads, for potential self-healing use in low-temperature marine environments, consisted of calcium alginate-encapsulated spores and mineral precursor compounds[238]. Their ability to form a biocomposite in simulated marine concrete crack solution at 8C was determined. Following 6 days of immersion in the solution, the bacteria-based beads had a calcite crust on their surfaces as well as calcite inclusions which resulted in a calcite-alginate biocomposite. The beads swelled by 300%, giving a maximum diameter of 3mm. Theoretical calculations predicted that 0.112g of beads could generate some $1mm^3$ of calcite in 14 days of immersion.

A strategy was developed[239] for the determination of the calcium precipitation activity of the mineralization bacteria used for self-healing. A strain (H4) having a calcium precipitation activity of 94.8% was identified and the effects of various factors were evaluated. It was found that lactates and nitrates were the best sources of carbon and nitrogen, at optimum concentrations of about 25 and 18mM, respectively. The H4 strain was able to maintain a high calcium precipitation activity at pH values ranging from 9.5 to 11.0, with an initial spore concentration of $4.0 \times 10^7/ml$. An ambient Ca^{2+} concentration which was greater than 60mM had a very negative effect upon the calcium precipitation activity and upon the growth of H4. It was concluded that maintaining a low Ca^{2+} concentration was necessary for the microbial self-healing of concrete. A method was proposed[240] for the supply of oxygen to aid microbial calcium precipitation. A controlled-release tablet which contained a 9:1 ratio of CaO_2 and lactic acid could provide a stable oxygen supply and maintain the pH level at 9.5 to 11.0 for 45 days while in contact with water. In the presence of oxygen, the self-healing H4 spores germinated more effectively and maintained a high metabolic activity. The H4 vegetative cells produced 50% more calcium precipitation than that in the absence of an oxygen supply. Calcium precipitation monitoring showed that H4, in a binary self-healing system, precipitated 27.5mM of calcium in the presence of an oxygen supply, after 32 days. The dissolved oxygen concentration in the solution decreased from 15 to 4mg/l, while only 6.9mM of calcium precipitation occurred with no oxygen supply. In further work[241], an optimum yeast-extract concentration of 5g/l in the sporulation medium was determined and the effects of various factors upon the microbial calcium precipitation of H4 in the presence of an oxygen-releasing tablet were studied. This showed that CaO_2 is preferable to other oxides in improving calcium precipitation by H4. This strain can precipitate highly insoluble calcium at a CaO_2 dose of 7.5 to 12.5g/l, and the most suitable spore concentration is 6×10^8/ml when the viable spore ratio is about 50%. Lactate was again confirmed to be the best carbon source and nitrate to be the best nitrogen source for aerobic incubation.

A specific study, using water permeability tests, was made of the effect of a healing agent upon the recovery of liquid tightness following cracking, and exposure to water immersion or wet-dry cycling[242]. It was found that the compressive strength of mortar which contained lightweight aggregates was unaffected by the presence of a healing agent. The recovery of water tightness also did not differ greatly for specimens, with or without a healing agent, when continuously immersed in water. The recovery of water tightness markedly increased for specimens which contained a healing agent, as compared with specimens without it, during wet-dry cycling.

The effect of hydrogel-encapsulated bacterial spores has been investigated[243] by comparing the crack-closure behaviour, with and without the bio-hydrogels. This showed that specimens with bio-hydrogels exhibited a clearly improved healing efficiency, as compared with that for controls which contained the hydrogel alone. The healing ratios for specimens with bio-hydrogels ranged from 70 to 100% when the cracks were smaller than 0.3mm. This was more than 50% higher than the results for samples with pure hydrogel. The maximum crack-bridging was about 0.5mm after 7 days. Purely hydrogel-incorporate samples exhibited the healing of cracks of only about 0.18mm. The total volume ratio of healing products in specimens with bio-hydrogels amounted to 2.2%; about 60% higher than that (1.37%) for those with hydrogel alone. The water permeability was decreased by an average of 68%. Non-bacterial specimens had a maximum healed crack-width of less than 0.3mm and the average water permeability was

Materials Research Forum LLC

https://doi.org/10.21741/9781644901373

decreased by only 15 to 55%[244]. It was noted that the addition of hydrogels produced a more porous microstructure, due to a non-optimum compatibility between the hydrogel and concrete, and a higher porosity could provide more space for precipitation. The latest hydrogel is cross-linked by alginate, chitosan and calcium ions[245]. The addition of chitosan improved the swelling of calcium alginate, and an opposite pH-response to calcium alginate occurred when the chitosan content in solution reached 1.0%. When that amount of chitosan was added to hydrogel beads, there was a 10.28% increase in compressive strength and a 13.79% increase in flexural strength, as compared with control samples. A crack of 4cm in length, with a width of 1mm, could be healed after adding between 2.54×10^5 and $3.07 \times 10^5/cm^3$ of spores which were encapsulated in chitosan-free hydrogel.

The effectiveness of another healing system, this time involving calcium glutamate, was evaluated[246] by performing tests at the macroscale, in the form of flexural and ultrasonic pulse velocity measurements, and at the nanoscale, in the form of nano-indentation. In order to compare the self-healing strategy with traditional methods, an external repair having the same composition as the healing agent was applied manually following damage. Both the healing ratio and the recovery of mechanical properties following self-healing were lower than those after externally applied healing, but were higher than the control sample values. This suggested that self-healing had a relatively low healing effectiveness. Differences in the healing efficiency between differing methods could be partially explained by bacterial activity. Mainly spores were involved in both methods. In the case of self-healing, mineral precipitation could be triggered only after some spores had become activated by cracking, and the number of germinated spores was limited. On the other hand, sufficient spores were activated during externally applied healing and plentiful mineral products were present after precipitation. Differences in healing efficiency were also related to the nutrition supply. During external repair, an adequate amount of liquid medium was available to fill the crack. This was not possible in the case of self-healing because infeasible amounts of nutrient would have to be incorporated.

Two types of bacteria-based additive were developed[247] in order to improve the self-healing ability, as measured by the area repair-rate. The experimental results showed that both additives could produce a self-healing cementitious material. The bacteria were cultured in a liquid medium containing 5.0g/l of peptone and 3.0g/l of yeast extract at a pH of 7.0, and autoclaved (121C, 25min). The final concentration of spores in the suspension was $10^8/ml$, and powders were obtained by oven-drying. One additive consisted of calcium lactate plus spore powder. Its incorporation led to decreases of some 14.7, 6.8 and 0.1% in the compressive strength after 3, 7 and 28 days of curing, respectively. The other additive consisted of calcium formate and spore powder. Its incorporation led to decreases of about 1.6 and 2.2% in the compressive strength after 3 and 7 days of curing, respectively. There was however an 8.1% increase after 28 days of curing. The carbonation depths of the control specimen, of specimens with the first addition and of specimens with the second addition were 6.6, 7.0 and 6.5mm, respectively, after 3 days of accelerated carbonation.

The viability of encapsulated spores was investigated, and fracture of the microcapsules upon cracking was confirmed by means of scanning electron microscopy[248]. The self-healing ability was judged in terms of the crack-healing ratio and water permeability. This indicated that the healing-ratio in specimens with bio-microcapsules was higher (48 to 80%) than that (18 to 50%) in those without bacteria. The maximum crack width which could be healed in bacterial specimens was 970μm, as compared with 250μm in non-bacterial ones. The overall water permeability in bacterial samples was some 10 times lower than that in non-bacterial ones. Wet-dry cycling stimulated self-healing in specimens with encapsulated bacteria, but no healing at all was observed in specimens that were stored in a relative humidity of 95%. This proved that the presence of liquid water is an essential requirement for self-healing.

Table 15. Crack healing ability and compressive strength of concrete after 28 days

Sample	W_{max} (μm)	R_f (MPa)	R_{sh} (MPa)
reference	56	57.2	29.1
abiotic control	124	55.0	28.4
bacteria	111	54.9	25.5
bacteria plus nutrient	273	57.3	36.0

W_{max}: maximum width of crack which can be healed, R_f: strength at first loading, R_{sh}: strength after self-healing

Urea hydrolysis is considered to be the most effective route to microbially-induced $CaCO_3$ precipitation. Under suitable conditions of pH and temperature, the quantity of urea has a marked effect upon the rate of urea degradation and $CaCO_3$ precipitation. A bacteria-based self-healing system was developed[249] by loading healing agents in ceramsite carriers. The self-healing efficiency was evaluated in terms of crack closure, restoration of compressive strength and of capillary water absorption. The best healing effect was obtained when the bacteria and an organic nutrient were co-immobilized on carriers.

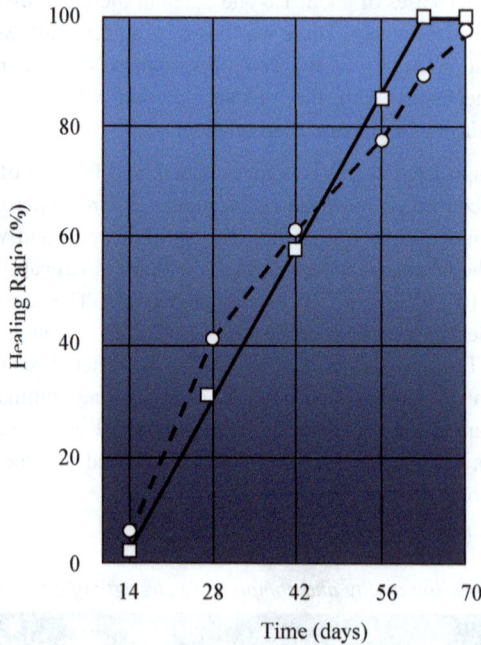

Figure 22. Crack healing of mortar containing
Lysinibacillus sphaericus bacteria. Squares: predicted, circles: measured

Cracks of up to 273µm could be healed, in 28 days, with a crack closure ratio of 86%. The restored compressive strength was increased by 24% (table 15), and the water absorption coefficient decreased by 27%, as compared to the control. No appreciable precipitation took place, in or on the rim of cracks, in the case of the control or other specimens. Bacteria- and nutrient-loaded specimens were quite different in that a large portion of the crack was filled with white precipitation products. The average and maximum widths of the cracks that could be healed were about 36 and 56µm for a mortar control sample. Organic matter could have a negative effect upon cement hydration and could retard the strength development of cementitious materials. No strength loss was found when organic substances were immobilized in carriers. A control series exhibited a 50% strength restoration following wet-dry cycling for 28 days, due to autogenous healing. The restoration ratio for abiotic control samples was almost the same as that for the control. This indicated that no chemical precipitation of a calcium source had contributed to healing. The strength restoration ratio attained more than 60% if bacteria and nutrients were loaded in ceramsite. This was expected because microbially-induced $CaCO_3$ precipitation occurred during the incubation period.

A lower restoration ratio of strength when compared to the reference was found when bacteria alone were loaded in ceramsite carriers. This implied an absence of bacteria-based healing. Calcite was a primary component in the other series. Calcite formation was attributed mainly to the metabolic activity of bacteria within the crack. This was also applicable to abiotic controls, due to the growth of other micro-organisms in the non-sterile immersion water. A calcium silicate hydrate gel phase and portlandite minerals in plate form predominated within the cracks in mortar control samples. Calcite appeared in the form of roughly 2μm rhombohedral grains in abiotic control series. Large $CaCO_3$ crystals, in the form of calcite, were found on the crack walls of bacteria-containing specimens. As compared with bacteria-loaded specimens, the calcite crystals seemed to be much more abundant in bacteria- and nutrient-loaded samples. This was suggested to explain the higher healing efficiency which resulted from simultaneously loading bacteria and nutrients in ceramsite.

Again using porous ceramsite, and considering the compatibility between healing agents and concrete, experiments[250] indicated that beef extract and peptone in self-healing agents had a negative effect while the effects of urea and calcium nitrate were not significant. The optimum volume ratio of ceramsite was 37.8%. Ceramsite-immobilized bacteria and organic nutrients were added to concrete. Following 28 days of curing, the recovery ratio of the compressive strength of cracked concrete attained about 63%. Its water-absorption was appreciably lower than that of control samples. The cracks were filled with calcite-type calcium carbonate, and the maximum crack width which could be healed was about 0.51mm. Sulfo-aluminate based expansive agents, crystalline admixtures and calcium hydrogen phosphate were used[251] as components of cementitious materials and porous ceramsites were used as carriers for sodium carbonate solution. Tests were performed on two preferred mixes in order to investigate changes in the gas-permeability of pre-cracked samples. Following curing in still water for 28 days, pre-cracked specimens with ceramsites containing sodium carbonate exhibited efficient surface-crack closure and healing of gas permeability. Further observations revealed formation of the healing product, $CaCO_3$, deeper within cracks of the chosen mix, as compared with a control mix.

In connection with the use of sodium carbonate, a type of cement-based healing pellet was proposed[252] which was based mainly on the concept that the introduced Na_2CO_3 would promote the formation of calcium carbonate in cracks. Higher contents of Na_2CO_3 decreased the size-range of pellets and reduced the setting time, fluidity and heat-of-hydration of the mortar. The pellets reduced the initial strength of the mortar, but maintained the intensity growth-rate stable so that there was hardly any negative effect upon the later strength. With increasing pellet content, the strength of the mortar decreased while the pore-filling efficiency and strength healing-rate of the mortar were improved. When the Na_2CO_3 content of the pellets, and the content of pellets in the mortar were 10 and 25%, respectively, the mortar attained its highest later-stage strength and exhibited the greatest healing effect. In other work lightweight clay aggregate, impregnated with liquid sodium carbonate, was used[253] for the autonomous self-healing of cementitious composites. It was found that, although the seepage of sodium carbonate markedly reduced the polarization resistance, the self-healing properties and the fresh and hardened properties of the concrete were improved. The healing products were present in both the interior, and at the mouth, of cracks and the crystals on the crack surface consisted mainly of calcite.

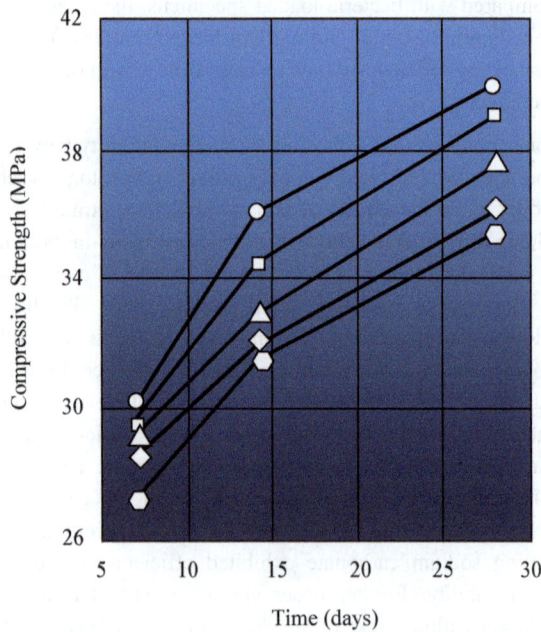

Figure 23. Compressive strength of concrete containing Lysinibacillus sphaericus bacteria. Circles: 2 x 10[7], squares: 10[7], triangles: 5 x 10[6], diamonds: control, hexagons: 10[8]

In other experiments[254], preliminary heat-treatment and NaOH-soaking were used to improve the loading content of ceramsite, and the viability of bacterial spores was checked by making urea-decomposition measurements. The results indicated that heat treatment could improve the loading content of ceramsite while not leading to a reduction in strength due to the ceramsite addition. The optimum heating temperature was 750C. The ceramsite particles protected spores from the high-pH environment of concrete. When nutrients and spores were both incorporated into ceramsite particles, the nutrients were easily available to the cells. The restoration-ratio of the compressive strength increased to over 20%, and the water-absorption ratio decreased by about 30% as compared with control samples. The healing-ratio of cracks attained 86%, and the maximum crack-width which could be healed was close to 0.3mm.

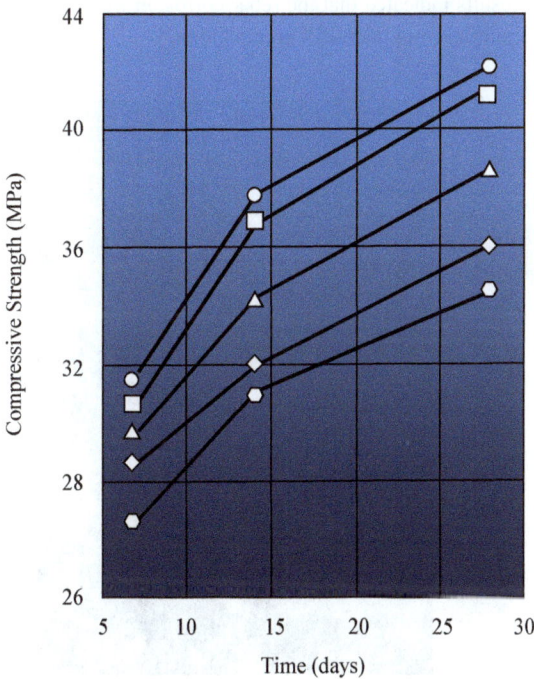

Figure 24. Compressive strength of concrete containing Sporosarcina pasteurii bacteria. Circles: 2 x 10[7], squares: 10[7], triangles: 5 x 10[6], diamonds: control, hexagons: 10[8]

In the most recent research[255] on ceramsite use, modified particles were used as carriers for bacteria by crushing and impregnation with the self-healing agent. Nutrients were embedded in the modified ceramsite in the form of powder. The particles were then encapsulated in an organic protective layer. The results showed that the immobilization ability, protective effect, mechanical properties and the interface of the ceramsite were markedly improved by modification. The crack-healing potential of concrete which contained ureolytic-type microbes, immobilized in porous ceramsite particles, was also studied[256] with regard to the resultant mechanical and electrochemical behaviours of the concrete. The maximum crack width which could be completely healed by the bacteria was up to 450µm within 120 days. The highest recovery ratios of the flexural strength and modulus were observed for bacterial concrete, for a given degree of initial damage. Tafel polarization results indicated that the rebar corrosion was successfully inhibited by the crack closure resulting from microbial precipitation.

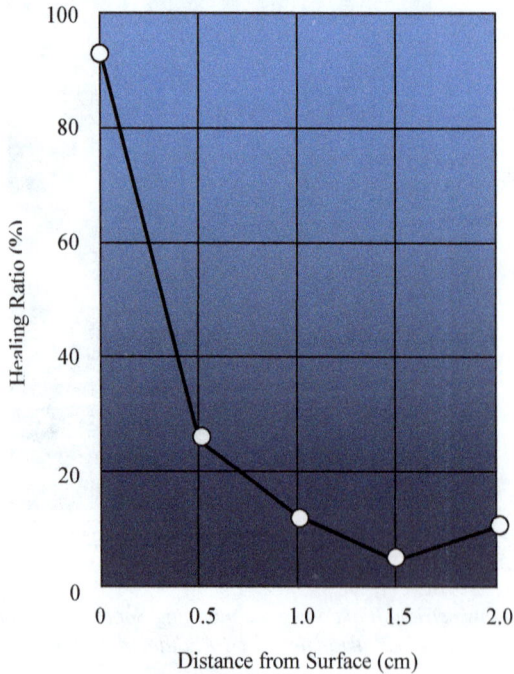

Figure 25. Crack healing by Lysinibacillus sphaericus
as a function of distance from the surface

Various single-species microbial cultures were coupled with protective carriers, showing that this combination could be useful in developing agents for creating self-healing concrete[257]. A mortar was studied which contained a healing agent that consisted of non-axenic biogranules with a de-nitrifying core. Specimens having a crack width of 400μm were treated with tap water for 28 days, and the self-healing was judged in terms of crack-volume reduction, thickness of the sealing layer along the crack depth and water permeability under a pressure of 0.1bar. During the first 14 days, the biomortar crack closure was not better than that of a control sample. All of the samples exhibited a crack-closure performance of between 10 and 30%. Complete observable crack closure was found for bio-treated specimens after 28 days and the thickness of the calcite layer was 10mm, with a healed-crack volume of 6%. The initial permeability of mortar with a 408μm wide crack was 29 x 10^{-5}m/s. A marked decrease in water permeability was found only in biomortars which contained active compact de-nitrifying core granules. Following the microbial-induced healing of 399μm cracks within 28 days, the permeability was 5 x 10^{-5}m/s; a 83% decrease in permeability. Calcium carbonate in the form of calcite, and aragonite, were the main constituents of the healing material. Precipitated calcium carbonate formed around bacterial clusters, and connected as the mineral grew.

A theoretical model has been developed[258] in order to simulate the kinetics of calcite-precipitation which results from the hydrolysis of urea in concrete cracks. A second-order partial differential equation was also derived which models the healing process on the basis of physical and biochemical factors. Scanning electron microscopic examination of artificially cracked mortar with incorporated *Lysinibacillus sphaericus* was used to check the predictions. The results (figure 22) suggested that prediction of the healing process of cementitious materials was possible by using these models. A related investigation was made[259] of bacterial growth and urea hydrolysis with the aim of promoting calcium carbonate precipitation using *Sporosarcina pasteurii* and *Lysinibacillus sphaericus* in environments having pH values of 7 to 13 and various concentrations of urea and calcium nitrate tetrahydrate. It was found that these bacteria could survive in a dormant state, with no reproduction, at pH values of 12 to 13. Urea hydrolysis was damped by the pH level, and its efficiency decreased by up to 75% within concrete pores at a pH value of 13. The best urea hydrolysis culture conditions were a pH level of 9, a concentration of calcium ions which was less than 150mM, an urea concentration of 333mM and a cell concentration of 2 x 10^8/ml. The compressive strength (figures 23 and 24) of concrete containing spores, vegetative cells and urea-vegetative cells was improved by 9, 10 and 15%, respectively, when compared with that of a control specimen. Complete healing, at the mouth, of a 0.4mm-wide crack occurred after 70 days. The healing ratio (figure 25), deeper in the concrete, was less than 15%. Although the strength increase was modest, these bacteria clearly had the ability to self-heal cracks in a concrete surface.

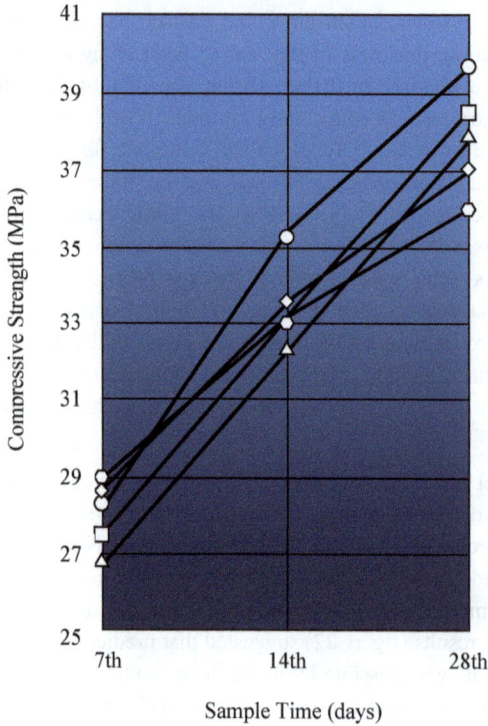

Figure 26. Compressive Strength of concrete with Enterococcus faecalis bacteria and calcium lactate. Circles: 2.18g/l, squares: 1.09g/l, triangles: 0.22g/l, diamonds: bacteria only, hexagons: control

The use of bacteria also offers an opportunity to clean up the environment. One innovation is to incorporate, as a fine aggregate in concrete, medical vial waste containing for example *Bacillus sphaericus*. The use of such waste also has economical advantages. The concept was tested[260] by using mixtures consisting of 15% medical vial waste and bacterial concentrations of 10^3, 10^5, 10^7, 10^9 or 10^{11} cells/ml of water. The aim was to find the mixture which led to the maximum load and maximum healing capacity. A mixture which consisted of 15% medical vial waste and a bacterial concentration of 10^7cells/ml of water satisfactorily improved the properties of concrete. Another proposed concept for cleaning up the environment, while simultaneously making economic gains, involves the biodegradable plastic, polyhydroxyalkanoate. This is habitually recovered from waste streams but is uneconomical to recycle. It can instead be used as a bacterial substrate in self-healing concrete. A proof-of-concept study[261] of the use of waste-derived polyhydroxyalkanoate as a bacterial substrate showed that such a polyhydroxyalkanoate-based healing system could induce crack-healing in concrete and reduce water permeability. The use of bacterial concrete may even help to counter global warming by sequestering CO_2. This involves the manufacture of bio-foamed concrete bricks while using *Bacillus tequilensis* to accelerate natural carbonation[262]. A suitable strain of *Bacillus tequilensis* was isolated from cement kiln dust. This produced carbon anhydrase and urease enzymes which accelerated the sequestration of CO_2. The sequestration of CO_2 in bio-foamed concrete bricks was optimised with regard to concrete density, *Bacillus tequilensis* concentration, temperature and CO_2 percentage. These factors had appreciable effects upon the sequestration process. The optimum carbonation depth of bio-foamed concrete bricks was 12.2mm after 28 days for 20% of CO_2, 3 x 10^5cells/ml of bacteria, a temperature of 40C and a density of 1300kg/m^3. The *Bacillus tequilensis* in bio-foamed concrete bricks increased the carbonation depth after 28 days by 30% as compared with foamed concrete bricks. The increase in carbonation depth resulted in increased formation of calcium carbonate in the bio-foamed concrete bricks as compared with foamed concrete bricks. Further environmental improvement would result from using biological additives in pavements. The use of native micro-organisms to produce healing products permits the autonomous self-healing of concrete pavements. Biocement production can source soil from local sites and add a mixture of nutrients and bacteria. As compared with synthetic topical sealers, biological surface sealers offer a more effective and safer means for avoiding the deterioration of concrete pavements[263].

The effects of *Enterococcus faecalis* and calcium lactate upon self-healing were studied[264] with respect to the compressive, flexural and tensile strengths. It was found that 2.18g/l of calcium lactate, plus 3% of the bacterium, increased the compressive strength from 36 to 39.6MPa and the flexural strength from 4.78 to 6.72MPa (figures 26 and 27). These increases were attributed to the calcium carbonate which formed due to the bacterium's ability to precipitate it via urease enzyme activity. The formation of calcium carbonate was also directly responsible for the self-healing capability (figure 28).

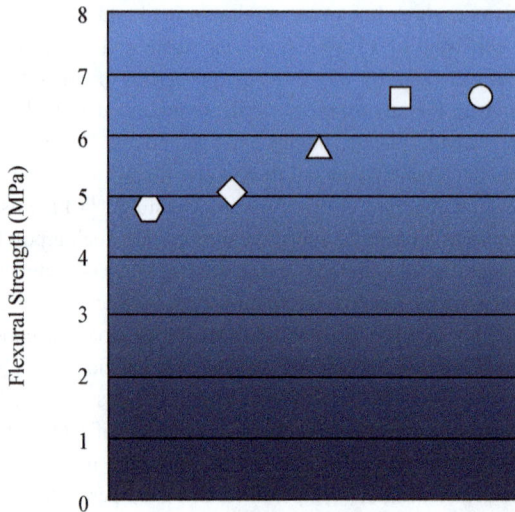

Figure 27. Flexural Strength of concrete with Enterococcus faecalis bacteria and calcium lactate. Circle: 2.18g/l, square: 1.09g/l, triangle: 0.22g/l, diamond: bacteria only, hexagon: control

A study was made[265] of the behaviour of a new isolate, *Bacillus specie* BY1, combined with organic calcium compounds such as calcium formate, calcium acetate and calcium lactate. The biominerals which formed in the presence of calcium formate and calcium lactate were mainly (about 95wt%) calcite, and had more rhombohedral faces, while calcium acetate produced spherulitic biominerals with a smaller (about 61.5wt%) fraction of calcite. Upon adding bacteria and these calcium compounds, the compressive strength was decreased and increased, respectively. The loss of strength due to the bacteria was compensated when bacteria and the calcium compounds were both added. Self-healing of cracks was not produced by adding only bacteria, and was more marked when calcium lactate was used as a biomineral precursor. With increasing bacterial concentration (3.1 x 10^5, 3.1 x 10^6 and 3.1 x 10^7cfu/ml), the 28-day compressive strength was reduced by 5.8, 6.6 and 9.5%, respectively. The loss of strength was attributed to the biomass behaviour in the cement matrix: the bacterial cells that could not sporulate then died during curing and their disintegration made the cement matrix more porous, and weaker. Calcium formate, calcium acetate and calcium lactate were added in the absence of the bacterium, and this also affected the compressive strength. That of calcium-containing specimens was lower than the strength of specimens without the organic compounds until after 3 days of curing. Their strength greatly increased following between 4 and 7 days of curing. Upon adding calcium formate, calcium acetate and calcium lactate, the 28-day compressive strength became 9.5, 9.9 and 16.4% higher, respectively. The increase in strength arose from the activity of the organic calcium compounds, which acted as cement-hydration accelerators.

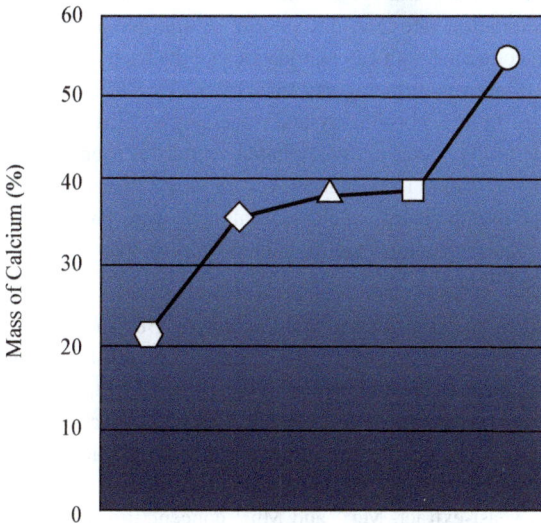

Figure 28. Calcium mass in various samples. Circle: 2.18g/l, square: 1.09g/l, triangle: 0.22g/l, diamond: bacteria only, hexagon: control

When both the bacterium and the calcium compounds were added, the latter still acted as hydration-accelerators and increased the strength. For calcium formate, calcium acetate and calcium lactate, the 28-day compressive strengths were 11.9, 15 and 11.5% higher, respectively. The self-healing of concrete microcracks of up to 100μm in width was monitored for 15 weeks after adding both bacteria and calcium compounds. When neither healing agent was incorporated, the microcracks remained unhealed. This also occurred when the bacterium alone was added. This indicated that bacteria alone were insufficient to form biominerals and fill the cracks. The addition of calcium sources to the bacteria produced differing self-healing behaviours, depending upon the type of calcium compound. The addition of calcium formate unexpectedly did not lead to self-healing of the microcracks, and a deficit in carbon mass required for biomineral formation was suggested to be the reason. The calcium compounds were added as 1% by weight of cement, generating carbon ratios of 1.107, 1.821 and 1.980 for calcium formate, calcium acetate and calcium lactate, respectively. The addition of calcium acetate or calcium lactate induced biomineral precipitation, with white precipitates filling microcracks at the specimen surfaces. A lower rate of self-healing was attributed to the polymorphic form of the biominerals. Rhombohedral calcite which formed in the presence of calcium lactate had a lower precipitation rate than did vaterite or amorphous $CaCO_3$. After 15 weeks, specimens which contained both calcium lactate and the bacterium exhibited marked self-healing activity; effectively effacing any trace of cracks. The higher self-healing activity could be explained by the morphology and mineralogy of the biominerals. The addition of calcium lactate to the bacterium precipitated crystals of rhombohedral shape having a higher calcite/vaterite ratio. These particles were larger than when calcium acetate was added. The larger rhombohedral calcite crystals that were produced by the addition of calcium lactate could promote the biomineralization that was associated with the pronounced crack-filling.

A study of another bacterial strain, *Bacillus specie* B8, showed[266] that the optimum concrete dose was 10^7/ml, with carbon and nitrogen nutrients being added in the form of starch and yeast extract at concentrations of 0.5 and 2g/l, respectively. The presence of Mg^{2+} and Mn^{2+}, in concentrations of 0.24 and 4.8mg/l, respectively, was also beneficial to spore formation. The inoculation volume and yeast-extract concentration were the most significant factors which affected spore yield. Under optimum conditions, with inoculate, starch, yeast-extract, Mg^{2+} and Mn^{2+} concentrations of 10^7/ml, 5, 5, 0.48 and 0.0048g/l, respectively, the final spore concentration attained 1.54×10^9/ml; a more than 150-fold increase. The spores required some 4h in order to complete germination. The most suitable germinant was found to be inosine at an optimum concentration of 2g/l and a preferred pH of 10. The presence of Na^+ could promote spore germination, while Ca^{2+} exerted a marked inhibitive effect. On the other hand, the presence of Na^+ in suitable concentrations (e.g. 24g/l) could largely counteract the negative effect of Ca^{2+}.

An alkali-tolerant bacterium, *Bacillus specie* AK13, has been isolated[267] from the rhizosphere of *Miscanthus sacchariflorus* (Amur Silvergrass). This strain can survive pH levels of 13, and brine containing 11%(w/v) of NaCl. Urea-free growth of the bacterium at high pH levels promoted the formation of calcium carbonate. Irregular vaterite-like $CaCO_3$ was observed which was tightly attached to cells. Most of the CO_3^{2-} ions which made up the $CaCO_3$ were produced by cellular respiration rather than coming from atmospheric carbon dioxide. The minerals which were produced from a growth medium which contained calcium acetate formed smaller crystals than those which appeared in media which contained calcium lactate. The strain appeared to heal cracks in mortar specimens when applied as a pelletized spore powder.

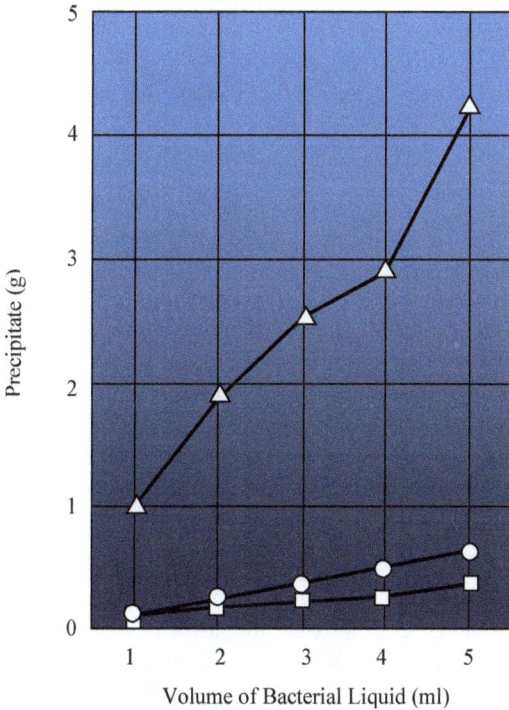

Figure 29. Quantity of precipitate produced by bacteria at various concentrations. Triangles: 10^9 cells/ml, circles. 10^8 cells/ml, squares: 10^7 cells/ml

Lysinibacillus boronitolerans YS11 was also isolated[268] from the rhizosphere of *Miscanthus sacchariflorus*. An increased pH level under the urea-free conditions during growth of the YS11 strain promoted the formation of calcium carbonate. Malt powder, rice bran, $(NH_4)_2SO_4$ and corn syrup were used to promote YS11 growth. Additional factors, such as the Mn^{2+}, Fe^{2+} and Ca^{2+} contributed to the essential process of sporulation. A mixture of *Lysinibacillus boronitolerans* YS11 spore powder, cement, sand, yeast extract, calcium lactate and water had a healing effect upon 0.3mm mortar cracks within 7 days and calcium carbonate precipitation occurred over the crack surface. Expanded clay was used[269] as a carrier to protect *Lysinibacillus boronitolerans* YS11 from the concrete environment during mixing. When exposed to a bacterial solution of 1.0×10^9cells/ml, the bacterial density within the expanded clay reached about 0.82×10^7cells/g of dry expanded clay. The extent of bacterial viability within clay submerged in a solution containing 1.0×10^8cells/ml, was 53.6% of free bacteria solution containing 1.0×10^7cells/ml. The rate of calcium carbonate formation was deduced from Ca^{2+} disappearance, and the rates were comparable for bacteria within expanded clay submerged in a bacterial solution containing 1.0×10^8cells/ml and for 1.0×10^7cells/ml of free bacteria. This indicated that bacteria in expanded clay are very active in the generation of $CaCO_3$ within the clay.

The ability of urobacteria to precipitate calcium carbonate is useful in the manufacture of self-healing concrete, and a hypersaline environment is an ecological niche for micro-organisms which are resistant to increased alkalinity and environmental stress. A study[270] of micro-organisms led to isolation of the most active urobacteria, *Lysinibacillus macroides* and *Bacillus licheniformis*, from hypersaline lakes. The introduction of these micro-organisms into the cement mixture markedly increased the strength of mortar specimens and reduced their porosity and capillary water absorption. This was associated with the progress of biocalcination. The microstructure of spores of diatomite-immobilized bacteria showed that this form permitted the long-term preservation of bacterial activity.

The $CaCO_3$-precipitation ability of ureolytic and non-ureolytic bacteria which were co-cultured as a self-healing agent for cementitious crack repair was determined[271]. Three different ratios (10:0, 8:2, 5:5) of ureolytic *Sporosarcina pasteurii* to non-ureolytic *Bacillus thuringiensis* were used. The effect of co-culturing ureolytic and non-ureolytic bacteria was investigated by measuring the rate of growth in urea-containing media, and the rates of NH_4^+ and $CaCO_3$ production in urea plus calcium lactate media. The self-healing ability of the co-cultured bacteria was judged by exposing specimens, which contained pre-defined cracks, to media that contained single urease-producing or co-cultured bacteria. The $CaCO_3$ precipitation was improved, by co-culturing ureolytic and non-ureolytic bacteria, due to the relatively faster growth-rate of the non-ureolytic bacteria. Specimens which contained narrower cracks exhibited a faster filling effect. This indicated that the crack width could be a predominant factor in influencing the $CaCO_3$-precipitation ability of co-cultured bacteria.

Materials Research Forum LLC
https://doi.org/10.21741/9781644901373

The urease-producing bacterium, *Bacillus pasteurii*, has been used[272] to investigate the relationship between the amount of substrate and the bacterial concentration by using samples with 10^7, 10^8 or 10^9cells/ml (figure 29). The healing effect was judged in terms of the area self-healing percentage and the impermeability ratio. For all three concentrations, the amount of substrate which was required increased linearly with increasing volume of bacterial liquid. For cracks having widths of 0.2 to 0.3mm, the healing efficiency at concentrations of 10^8 and 10^9/ml was much higher than that at 10^7/ml. The area self-healing percentages were 89.38 94.56 and 68.21%, and the impermeability ratios were 70.13, 86.52 and 53.69%, respectively, for bacterial concentrations of 10^8, 10^9 and 10^7/ml. For cracks having widths of 0.6 to 0.7mm, the healing efficiency was much lower than that for small cracks. The self-healing percentage and the impermeability ratio increased with increasing bacterial concentration. The area self-healing percentages were 29.68, 48.39 and 75.39%, and the impermeability ratios were 25.36, 39.67 and 51.73%, respectively, for bacterial concentrations of 10^7, 10^8 and 10^9/ml. Due to the production of calcium carbonate during self-healing, mortar permeability can be reduced. Specimens were cast using *Sporosarcina pasteurii* bacteria, and held in urea plus calcium chloride or urea plus calcium lactate curing environments[273]. This led to an increased compressive strength and electrical resistance. The greatest improvement, of 60% in the 28-day compressive strength, was found for specimens which contained bacteria and were cured in the urea plus calcium chloride solution. The water-absorption of the self-healing mortar decreased by 32 to 55%, relative to that of control samples. The urea plus calcium lactate curing environment was more effective than the urea plus calcium chloride one in providing calcium ions for the formation of calcium carbonate sediments, such that water-absorption was reduced by 49 and 55% when specimens were cured in the urea plus calcium lactate medium. Alternative curing environments had no appreciable effect upon the compressive strength of specimens with essentially identical amounts of precipitated calcium carbonate crystals. The use of biological self-healing led to an increased resistance to chloride-ion penetration. Specimens which contained bacteria and calcium carbonate passed reduced amounts of electrical charge. A greater decrease was observed for chloride-ion permeability in specimens which were cured for 28 days in the urea plus calcium lactate environment. This was attributed to the fact that this environment produced more calcium carbonate. In another study[274] of the effect of calcium lactate and *Bacillus subtilis*, there was a decrease in the carbonation rate and an improvement in the compressive strength. Scanning electron microscopy showed that the matrix density of the bacterial concrete was greater than that of a control mix.

Lightweight aggregate was impregnated with a solution of *Bacillus pseudofirmus* B-4104 for use as a self-healing system[275]. The system could produce a 112% recovery of the splitting tensile strength, a 82% decrease in the gas permeability and a 87.4% decrease in water sorptivity. Crystalline products which were compatible with the cement matrix filled the microcracks at the nanoscale and eventually healed the samples. Polyurea polymer was proposed[276] as a carrier for *Bacillus pseudofirmus* so that it could produce calcium carbonate crystals. Encapsulation of the bacteria in the polyurea was achieved by *in situ* polymerization. The bacterial spores, and nutrition, could be encapsulated without affecting their chemical structure. The polyurea microcapsules which contained the bacteria were mixed with cement paste. The hardened specimens were cracked by using a 3-point bending apparatus in order to trigger artificially the healing effect of the bacterial capsules. Further study confirmed the precipitation of calcium carbonate around the crack-zone.

A ratiometric pH optode imaging system, with a resolution of $50\mu m$/pixel, was used[277] to characterize the pH of the pore-water within the cracks of submerged hydrated cement; with an accuracy of 1.4pH-units/mm. The pH decreased markedly, from more than 11 to below 10, with increasing fly-ash content and hydration time. Bio-activity in the cement was assessed by introducing super-absorbent polymers, with encapsulated *Bacillus alkalinitrilicus* endospores, into the cracks. The bacterial activity was measured by using oxygen optodes, showing that the greatest bacterial activity occurred with increasing amounts of fly-ash substitution in the cement.

Bacillus cereus at a concentration of 10^7cells/ml, was used[278] to study the production of $CaCO_3$ crystals in the presence of 20g of Portland cement and 0.58, 1.00 or 1.42g of calcium lactate. The samples were held at 28C for 168h. After 17h, crystals had started to appear in samples which contained 1.00 or 1.42g of calcium lactate. After 168h, such crystals were present in all of the samples. In the sample with the highest lactate content, there was 82% SiO_2 and 18% $CaCO_3$.

A biomimetic strategy has been used[279] to improve the self-healing of Portland cement by using biogenic inorganic polymer polyphosphate (polyP) as a cement admixture. Synthetic linear polyP, with an average chain-length of 40, as well as natural long-chain polyP isolated from soil bacteria, could support self-healing of this material. The polyP, when used as a water-soluble Na-salt, underwent Na^+/Ca^{2+} exchange with Ca^{2+} from the cement, resulting in the formation of a water-rich coacervate when added to the cement surface, and especially to the surface of bacteria-containing cement/concrete samples. The addition of polyP at low concentrations (<1wt%) accelerated hardening of the cement and the healing of microcracks. The results suggested that long-chain polyP was a promising additive for increasing the self-healing ability by mimicking a bacteria-mediated mechanism.

Alkali-resistant spore-forming bacteria which were related to the genus *Bacillus* were selected for a study[280] in which the performance of the bio-based agent was assessed by monitoring the mechanical properties and durability of bio-remediated fly-ash concrete. Various amounts of bacterial cell concentrations were introduced into artificially created cracks, showing that there was an up to 15.6% increase in the compressive strength upon adding 10^5cells/ml. Because the calcite which was formed then filled up the pores, this bio-based agent could potentially decrease the porosity and increase the strength and durability of fly-ash concrete. The effect was studied[281] of alkaliphilic spore-forming bacteria of the genus *Bacillus* upon the compressive strength of mortar cubes and the healing ability of the bacteria. Cracked specimens, with and without the bacteria, were prepared showing that inclusion of the bio-healing agent led to a greater pore volume which had no appreciable effect upon the compressive strength after 28 days. Cracks were more significantly healed in bacteria-based, than in control, specimens after 28 days of incubation in a water bath. This increased the permeability resistance of bacteria-based specimens.

Studies of bacteria-based self-healing concrete show that it is necessary to encapsulate and separate the self-healing ingredients (bacteria, nutrients, precursors) so that, when a crack forms, the capsules rupture and permit the self-healing ingredients to come together and precipitate calcite in the crack. Because of the shearing action of concrete mixers, there is a chance that the capsules may break prematurely; affecting the efficiency of self-healing and detrimentally affecting the concrete's properties. Tests were made[282] of the effects of multi-component growth media, containing germination and sporulation aids for the bacterial aerobic oxidation, upon the properties of fresh and hardened concrete. The results showed that a multi-constituent growth medium would not have a great effect upon the properties of concrete in the proportions which were likely to be released during mixing.

An attempt[283] was made to correlate microbial-induced $CaCO_3$ precipitation, based upon enzymatic urea hydrolysis, with the optimum requirements of the self-healing of cracks in concrete. Experiment showed that the initial cell density and Ca^{2+} concentration were significant factors. A high initial cell density (10^8/cells/ml) and a relatively low (50mM) Ca^{2+} concentration favoured microbial precipitation. A secondary study was made of the dissolution behaviours of urea and calcium, given that the solution of healing agents in cracks was required for self-healing. Upon adding urea and $Ca(NO_3)_2$ to concrete in a constant mass ratio of 2:3, the highest estimated values of urea content (345mM) and Ca^{2+} concentration (44mM) in cracks were close to the optimum values suggested by experiment. Although the addition of urea and $Ca(NO_3)_2$ would not have a negative effect upon the mechanical properties of concrete, direct mixing was not advised.

Recent research[284] has involved the development of bio concrete and autonomous crack-healing in submerged marine structures, especially with regard to the optimum nutrient precursor for the bio-concrete. Calcium lactate was found to be the optimum nutrient because it increased, by 16%, the overall healing under submerged marine conditions and improved the aragonite/brucite ratio; a higher ratio being preferred because aragonite is harder than brucite. Bio-concrete samples were prepared using Halobacillus *halophilus* bacteria, with expanded perlite aggregate as the carrier, plus calcium lactate. This improved autonomous crack healing by 17% under both submerged and tidal marine conditions, with the healing products being visible along the entire crack depth under the submerged conditions. The bacteria were able to increase the aragonite/brucite ratio, throughout the crack depth, due to the availability of water and oxygen within the cracks.

In a general study of the too-shallow repair of microbial self-repairing cement-based materials, the effects of crack width, cracking age and crack depth upon the physicochemical characteristics of the crack-mouth solution were investigated[285]. This showed that, in the closed environment, the pH value in the crack-mouth solution was about 13.3. When the cracking age of the specimen was increased from 3 to 56 days, the Ca^{2+} concentration in the crack-mouth solution gradually increased from 140 to 350mg/l. In the atmospheric environment, the pH value of the surface layer 3mm crack-mouth solution was relatively low, gradually increased along the depth of the crack and then stabilized at about 13.2. The Ca^{2+} concentration in the crack mouth gradually increased in the depth direction and tended to be stable. The CO_3^{2-} concentration in the crack-mouth solution gradually decreased in the depth direction and tended to stabilize at about 1.5g/l. Following the addition of micro-organisms to the crack-mouth solution, the CO_3^{2-} concentration in the surface layer increased while the CO_3^{2-} concentration in the depth direction did not appreciably increase. The lower carbonate-ion concentration in the depth of a crack was the main reason for too-shallow repairing of the crack mouth. The mineralized product was mainly calcium carbonate of calcite-type in the crack-mouth solution.

Fungi

A recent innovation has been the suggested exploitation of fungi as self-healing agents[286]. The concept is that the fungi will promote the precipitation of calcium-based minerals which then fill cracks in the concrete (figure 30). It is important of course that they should be non-pathogenic to animals, and this is fortunately true of some 96.6% of them (the meat-substitute, Quorn, is made from such fungi). They in turn must be able to survive the mixing process as well as the strongly alkaline environment (with a typical pH value of 13) of fresh concrete, created by the formation of $Ca(OH)_2$: *Paecillomyces lilacimus* can thrive when the pH is between 7.5 and 11.0. The most promising fungal candidates are therefore alkaliphilic spore-forming ones, so that their spores and nutrients can be disseminated during mixing. Given sufficient water and oxygen, dormant spores will grow and precipitate $CaCO_3$ when needed, thus filling any cracks. When the cracks are completely filled and no more water or oxygen can enter, the fungi will again form spores. It might even be possible to manipulate the fungus genetically so as to tailor it to the purpose of crack repair. They can already survive in arctic and desert environments and so should be able to protect any concrete structures in those locations. It is expected that some species of fungi can promote calcium mineralization. The existence of calcified fungal traces in calcareous soils and limestone implies that fungi can play a role in secondary $CaCO_3$ precipitation. The formation of calcium minerals by *Serpula himantioides* has been observed. The use of fungi in biogenic crack repair is expected to be more effective than bacterial use due to their great ability to promote calcium mineralization. The mechanism of calcium mineralization by fungi is not entirely understood, but appears to involve cation binding to fungal-cell walls and the formation of calcite via the fungal excretion of hydrogen ions or organic acids. Cation binding results in mineral nucleation and deposition, and bound calcium cations interact with soluble CO_3^{2-}, leading to $CaCO_3$ deposition on fungal filaments. In one study[287], a fungal growth medium was overlaid onto cured concrete and mycelial discs were aseptically deposited. Experimental results showed that, due to the dissolution of $Ca(OH)_2$ from concrete, the pH of the growth medium increased from 6.5 to 13.0. Spores of *Trichoderma reesei* could nevertheless germinate into mycelium and grew equally well, on or off the concrete. Others did not grow on the concrete. Crystals which precipitated on the fungus were found to be composed of calcite. Other research[288] showed that *Aspergillus nidulans*, a pH-regulatory mutant, could grow on concrete and promote calcium carbonate precipitation. The growth of various fungi was compared by using a layer of potato dextrose agar as growth medium, with or without concrete or a buffer to control acidity (table 16).

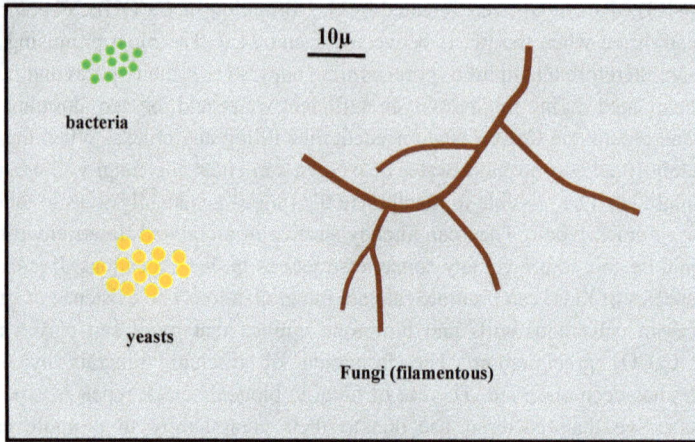

Figure 30. Morphological and size comparison of filamentous fungi, yeast and bacteria; suggesting the fungal branching behaviour can provide copious nucleation sites and mechanical support for calcium carbonate precipitates

Cation binding by fungi, a metabolism-independent process, binds ions to cell-walls and similar surfaces and leads to mineral nucleation and deposition. Bound calcium cations tend to interact with soluble CO_3^{2-} ions, thus leading to the deposition of $CaCO_3$ on the filaments of the fungus. Even dead and metabolically inactive fungal components act as nucleation sites for additional calcium carbonate precipitation. The presence of chitin, a modified nitrogen-containing polysaccharide, markedly lowers the activation energy for nucleus formation so that calcite can easily nucleate and grow on it. The excretion of organic acids, such as oxalic, by the fungal filaments also plays an important role in the re-precipitation of secondary calcium phases in the $CaCO_3$-rich environment. The oxalic acid which is produced by *Aspergillus niger* can react with $CaCO_3$ so as to form calcium oxalate. The excretion of oxalic acid and the precipitation of calcium oxalate can then lead to dissolution of the internal pore walls of a limestone matrix and enrich the solution in CO_3^{2-}.

Table 16. Growth rate (mm/day), 21 days following inoculation,
of fungi on potato dextrose agar

Species	Substrate	Rate
Trichoderma reesei	agar incubated at 30C	2.6
Trichoderma reesei	agar incubated at 25C	2.6
Trichoderma reesei	agar with buffer* incubated at 30C	1.0
Trichoderma reesei	agar with buffer incubated at 25C	0.8
Trichoderma reesei	agar with concrete incubated at 30C	2.6
Trichoderma reesei	agar with concrete incubated at 25C	0
Trichoderma reesei	agar with concrete and buffer incubated at 30C	2.6
Trichoderma reesei	agar with concrete and buffer incubated at 25C	0
Aspergillus nidulans	agar incubated at 30C	2.6
Aspergillus nidulans	agar incubated at 25C	2.6
Aspergillus nidulans	agar with buffer incubated at 30C	0.5
Aspergillus nidulans	agar with buffer incubated at 25C	0.7
Aspergillus nidulans	agar with concrete incubated at 30C	0
Aspergillus nidulans	agar with concrete incubated at 25C	0
Aspergillus nidulans	agar with concrete and buffer incubated at 30C	0
Aspergillus nidulans	agar with concrete and buffer incubated at 25C	0
Cadophora interclivum	agar incubated at 30C	0.6
Cadophora interclivum	agar incubated at 25C	2.1
Cadophora interclivum	agar with buffer incubated at 30C	0
Cadophora interclivum	agar with buffer incubated at 25C	0.6
Cadophora interclivum	agar with concrete incubated at 30C	0
Cadophora interclivum	agar with concrete incubated at 25C	0
Cadophora interclivum	agar with concrete and buffer incubated at 30C	0
Cadophora interclivum	agar with concrete and buffer incubated at 25C	0
Umbeliopsis dimorpha	agar incubated at 30C	2.6

Umbeliopsis dimorpha	agar incubated at 25C	2.6
Umbeliopsis dimorpha	agar with buffer incubated at 30C	1.0
Umbeliopsis dimorpha	agar with buffer incubated at 25C	1.0
Umbeliopsis dimorpha	agar with concrete incubated at 30C	0
Umbeliopsis dimorpha	agar with concrete incubated at 25C	0
Umbeliopsis dimorpha	agar with concrete and buffer incubated at 30C	0
Umbeliopsis dimorpha	agar with concrete and buffer incubated at 25C	0
Acidomelania panicicola	agar incubated at 30C	2.1
Acidomelania panicicola	agar incubated at 25C	1.9
Acidomelania panicicola	agar with buffer incubated at 30C	0.9
Acidomelania panicicola	agar with buffer incubated at 25C	0.9
Acidomelania panicicola	agar with concrete incubated at 30C	0
Acidomelania panicicola	agar with concrete incubated at 25C	0
Acidomelania panicicola	agar with concrete and buffer incubated at 30C	0
Acidomelania panicicola	agar with concrete and buffer incubated at 25C	0
Pseudophialophora magnispora	agar incubated at 30C	2.6
Pseudophialophora magnispora	agar incubated at 25C	2.1
Pseudophialophora magnispora	agar with buffer incubated at 30C	0
Pseudophialophora magnispora	agar with buffer incubated at 25C	0
Pseudophialophora magnispora	agar with concrete incubated at 30C	0
Pseudophialophora magnispora	agar with concrete incubated at 25C	0
Pseudophialophora magnispora	agar with concrete and buffer incubated at 30C	0
Pseudophialophora magnispora	agar with concrete and buffer incubated at 25C	0

* buffered with 3-(N-morpholino)propanesulfonic acid

As the solution passes through the pore walls, $CaCO_3$ can recrystallize due to the decreased CO_2 level. Biodegradation of an oxalate by microbial activity can also result in transformation to CO_3^{2-}, leading to $CaCO_3$ precipitation within pores. During the decomposition of fungal filaments, released $CaCO_3$ crystals could act as secondary calcite-precipitation sites. The CO_2 production which results from both oxalate oxidation and fungal respiration can cause the concentration of CO_3^{2-} in the local environment and promote more $CaCO_3$ precipitation.

The most recent studies[289] show that *Fusarium oxysporum* is able to germinate on the surface of concrete and undergo rapid mycelium production, although the alkaline environment tends to slow their growth. Under favourable conditions the hydrophobic *Fusarium oxysporum* mycelium were nevertheless able to cover, within 96h, a surface area which was 86 times larger than the area of the original fungus deposit. Such rapid growth might clearly contribute to highly efficient crack-healing. The water-repellent nature of the mycelium could reduce the water-infiltration rate to 17.22% of that of the normal concrete surface.

About the Author

Dr. Fisher has wide knowledge and experience of the fields of engineering, metallurgy and solid-state physics, beginning with work at Rolls-Royce Aero Engines on turbine-blade research, related to the Concord supersonic passenger-aircraft project, which led to a BSc degree (1971) from the University of Wales. This was followed by theoretical and experimental work on the directional solidification of eutectic alloys having the ultimate aim of developing composite turbine blades. This work led to a doctoral degree (1978) from the Swiss Federal Institute of Technology (Lausanne). He then acted for many years as an editor of various academic journals, in particular *Defect and Diffusion Forum*. In recent years he has specialized in writing monographs which introduce readers to the most rapidly developing ideas in the fields of engineering, metallurgy and solid-state physics. He is co-author of the widely-cited student textbook, *Fundamentals of Solidification*, a new edition of which is in preparation. Google Scholar credits him with 7670 citations and a lifetime h-index of 16.

Keyword Index

3-aminopropyltriethoxy silane, 81

abiotic, 89-91
acacia-gum, 56
Acidomelania panicicola, 110
aerobic, 5, 71, 87, 105
aggregates, 1, 5, 8, 19, 23, 42, 48, 65, 75, 87
alkali-carbonate reaction, 6
alkaliphilic, 69, 75, 105, 107
aluminium stearate, 39
ammonification, 5
Amur Silvergrass, 100
anaerobic, 5, 53, 71, 76
Aspergillus nidulans, 107, 109
Aspergillus niger, 108
autogenous, 2, 19-22, 24, 32, 34-35, 41, 50, 75, 90
autolytic, 27

Bacillus alkalinitrilicus, 104
Bacillus anthracis, 84
Bacillus cohni, 69, 71, 82
Bacillus licheniformis, 102
Bacillus megaterium, 72
Bacillus pasteurii, 6, 40, 84, 86, 103
Bacillus pseudofirmus, 104
Bacillus sphaericus, 38, 75, 79, 82, 85, 96
Bacillus subtilis, 75-76, 78-85, 103
Bacillus tequilensis, 97
bentonite, 27, 35, 42, 79
biocalcination, 102
biocement, 66
biochar, 73-74
bio-hydrogel, 87
biomimetic, 53, 104
biomineralization, 6, 83, 85, 100
biomortar, 95
bioprecipitation, 85
bioremediase, 76
blast furnace slag, 25, 39, 47
brine-caused corrosion, 24
brucite, 22, 27, 52, 106

Cadophora interclivum, 109

calcite, 22, 28, 32, 35, 38, 48, 52, 61, 68, 70-71, 75-76, 83, 85-86, 91-92, 95, 98, 100, 105-108, 111
calcium aluminate phosphate, 24
calcium aluminium sulphate, 23
calcium carbonate, 4, 7, 13, 18, 25, 27, 34, 37-38, 40, 55, 60, 65-69, 71, 74-75, 79, 81-82, 84-86, 91, 95, 97, 100-104, 106-108
calcium formate, 88, 98-99
calcium hydroxide, 6, 28, 46
calcium ions, 5, 69, 88, 95, 103
calcium lactate pentahydrate, 82
calcium nitrate tetrahydrate, 95
calcium silicate hydrate, 12, 25-26, 34-36, 42, 55, 60, 83, 91
calcium sulfo-aluminate, 26, 33
carbide slag, 5
carbonate halo, 6
carbonic anhydrase, 78
cement, 1, 5-9, 12-13, 16, 18, 20, 22, 25-26, 28, 30, 32, 34-39, 46-47, 49, 52-56, 60-62, 64-66, 69, 71-72, 79, 81-82, 86, 90-91, 97-98, 100, 102, 104, 106
cementitious, 5-8, 11-12, 16, 18-20, 24-27, 32, 34, 36, 39, 44, 46-47, 53, 55-56, 59, 61, 71, 79, 88, 90-92, 95, 102
ceramsite, 89, 91, 93-94
chelation, 23, 25
chelator, 26
chitosan, 88
chloride-ion, 10, 12, 53, 72, 103
coacervate, 104
coir, 82
compressive strength, 5, 13, 19-20, 23, 25, 32-33, 36, 38-40, 44-45, 47-48, 50, 52, 55, 60-61, 63, 71, 73, 76, 80-85, 87-91, 93, 95, 97-99, 103, 105
crack-zone, 5, 104
curing, 2, 12, 16, 18, 24-25, 28, 31-32, 34, 36-39, 42, 45, 47-49, 54, 57, 59, 61, 67, 71-72, 79, 81, 83, 88, 91, 98, 103
cyanoacrylate, 47, 54
cyanobacteria, 40
debonding, 24

References

[1] Roig-Flores, M., Serna, P., Sustainability, 12[11] 2020, 4476. vanessa.coulet@univ-amu.fr

[2] Zhang, X., Qian, C., Smart Materials and Structures, 29[10] 2020, 105035. https://doi.org/10.1088/1361-665X/aba53b

[3] Wu, M., Hu, X., Zhang, Q., Cheng, W., Xue, D., Zhao, Y., Cement and Concrete Composites, 113, 2020, 103718. https://doi.org/10.1016/j.cemconcomp.2020.103718

[4] Han, S., Jang, I., Choi, E.K., Park, W., Yi, C., Chung, N., Journal of Environmental Engineering, 146[7] 2020, 04020072. https://doi.org/10.1061/(ASCE)EE.1943-7870.0001713

[5] Liu, C., Xu, X., Lv, Z., Xing, L., Journal of Advanced Concrete Technology, 18[4] 2020, 168-178. https://doi.org/10.3151/jact.18.168

[6] Liu, C., Lü, Z., Xiao, J., Bai, G., Journal of Building Materials, 23[6] 2020, 1337-1344.

[7] Chen, H.J., Chen, M.C., Tang, C.W., Sustainability, 12[3] 2020, 1242. https://doi.org/10.3390/su12031242

[8] Stukovnik, P., Bosiljkov, V.B., Marinsek, M., Materiali in Tehnologije, 54[3] 2020, 379-384. https://doi.org/10.17222/mit.2019.196

[9] Cai, X., Huang, W., Wu, K., Materials, 12[21] 2019, 3488. https://doi.org/10.3390/ma12213488

[10] Snoeck, D., Criel, P., Advances in Cement Research, 31[6] 2019, 261-269. https://doi.org/10.1680/jadcr.16.00091

[11] Zhu, H., Zhou, S., Yan, Z., Ju, W., Chen, Q., Computers and Concrete, 15[1] 2015, 37-54. https://doi.org/10.12989/cac.2015.15.1.037

[12] Aliko-Benítez, A., Doblaré, M., Sanz-Herrera, J.A., International Journal of Solids and Structures, 69-70, 2015, 392-402. https://doi.org/10.1016/j.ijsolstr.2015.05.011

[13] Quayum, M.S., Zhuang, X., Rabczuk, T., Journal of Contemporary Physics, 50[4] 2015, 383-396. https://doi.org/10.1007/s11709-015-0320-z

[14] Hazelwood, T., Jefferson, A.D., Lark, R.J., Gardner, D.R., Engineering Structures, 102, 2015, 176-188. https://doi.org/10.1016/j.engstruct.2015.07.056

[15] Suleiman, A.R., Nehdi, M.L., Materials, 10[2] 2017, 135. https://doi.org/10.3390/ma10020135

[16] Caggiano, A., Etse, G., Ferrara, L., Krelani, V., Computers and Structures, 186, 2017, 22-34. https://doi.org/10.1016/j.compstruc.2017.02.005

[17] Kazemi, A., Baghani, M., Shahsavari, H., Abrinia, K., Baniassadi, M., International Journal of Applied Mechanics, 9[6] 2017, 1750082. https://doi.org/10.1142/S175882511750082X

[18] Mauludin, L.M., Zhuang, X., Rabczuk, T., Composite Structures, 196, 2018, 63-75. https://doi.org/10.1016/j.compstruct.2018.04.066

[19] Oh, S.R., Choi, Y.W., Kim, Y.J., Construction and Building Materials, 214, 2019, 574-580. https://doi.org/10.1016/j.conbuildmat.2019.04.123

[20] Oucif, C., Voyiadjis, G.Z., Rabczuk, T., Theoretical and Applied Fracture Mechanics, 96, 2018, 216-230. https://doi.org/10.1016/j.tafmec.2018.04.010

[21] Kazemi, A., Baghani, M., Shahsavari, H., Sohrabpour, S., International Journal of Applied Mechanics, 10[7] 2018, 1850077. https://doi.org/10.1142/S1758825118500771

[22] Xu, J., Peng, C., Wan, L., Wu, Q., She, W., Journal of Materials in Civil Engineering, 32[6] 2020, 04020149. https://doi.org/10.1061/(ASCE)MT.1943-5533.0003214

[23] Mauludin, L.M., Budiman, B.A., Santosa, S.P., Zhuang, X., Rabczuk, T., Journal of Mechanical Science and Technology, 34[5] 2020, 1847-1853. https://doi.org/10.1007/s12206-020-0405-z

[24] Ma, H., Herbert, E., Ohno, M., Li, V.C., Cement and Concrete Composites, 95, 2019, 1-9. https://doi.org/10.1016/j.cemconcomp.2018.10.006

[25] Xue, C., Li, W., Li, J., Wang, K., Cement and Concrete Research, 122, 2019, 1-16. https://doi.org/10.1016/j.cemconres.2019.04.012

[26] Zhang, Z., Zhang, Q., Li, V.C., Cement and Concrete Composites, 103, 2019, 293-302. https://doi.org/10.1016/j.cemconcomp.2019.05.014

[27] Jiang, J., Zheng, X., Wu, S., Liu, Z., Zheng, Q., Construction and Building Materials, 198, 2019, 696-709. https://doi.org/10.1016/j.conbuildmat.2018.11.054

[28] Chen, J., Ye, G., Cement and Concrete Research, 123, 2019, 105782. https://doi.org/10.1016/j.cemconres.2019.105782

[29] Wang, X., Zhang, J., Han, R., Han, N., Xing, F., Journal of Cleaner Production, 235, 2019, 966-976. https://doi.org/10.1016/j.jclepro.2019.06.294

[30] He, H., Zhu, Y., Zhou, A., Construction and Building Materials, 188, 2018, 153-160. https://doi.org/10.1016/j.conbuildmat.2018.08.104

[31] Yuan, Z.C., Jiang, Z.W., Chen, Q., Journal of Central South University, 26[3] 2019, 567-576. https://doi.org/10.1007/s11771-019-4028-4

[32] Pan, Y., Tian, F., Zhong, Z., International Journal of Damage Mechanics, 27[5] 2018, 754-778. https://doi.org/10.1177/1056789517702211

[33] Jacobsen, S., Sellevold, E.J., Cement and Concrete Research, 26[1] 1996, 55-62. https://doi.org/10.1016/0008-8846(95)00179-4

[34] Mihashi, H., Kaneko, Y., Nishiwaki, T., Otsuka, K., Transactions of the Japan Concrete Institute, 22, 2000, 441-450.

[35] Reinhardt, H.W., Jooss, M., Cement and Concrete Research, 33[7] 2003, 981-985. https://doi.org/10.1016/S0008-8846(02)01099-2

[36] Liu, X.Y., Yao, W., Zheng, X.F., Wu, J.P., Journal of Building Materials, 8[2] 2005, 184-188.

[37] Yao, W., Zhong, W., Chinese Journal of Materials Research, 20[1] 2006, 24-28. https://doi.org/10.1142/S0256767906001047

[38] Zhong, W., Yao, W., Construction and Building Materials, 22[6] 2008, 1137-1142. https://doi.org/10.1016/j.conbuildmat.2007.02.006

[39] Nishiwaki, T., Mihashi, H., Jang, B.K., Miura, K., Journal of Advanced Concrete Technology, 4[2] 2006, 267-275. https://doi.org/10.3151/jact.4.267

[40] Şahmaran, M., Keskin, S.B., Ozerkan, G., Yaman, I.O., Cement and Concrete Composites, 30[10] 2008, 872-879. https://doi.org/10.1016/j.cemconcomp.2008.07.001

[41] Thao, T.D.P., Johnson, T.J.S., Tong, Q.S., Dai, P.S., IES Journal A, 2[2] 2009, 116-125. https://doi.org/10.1080/19373260902843506

[42] Li, V., Materials Performance, 48[6] 2009, 20-22.

[43] He, H., Guo, Z., Stroeven, P., Stroeven, M., International Journal of Modelling, Identification and Control, 7[2] 2009, 142-147. https://doi.org/10.1504/IJMIC.2009.027067

[44] Li, V.C., Herbert, E., Journal of Advanced Concrete Technology, 10[6] 2012, 207-218. https://doi.org/10.3151/jact.10.207

[45] Abd Elhakam, A., Mohamed, A.E., Awad, E., Construction and Building Materials, 35, 2012, 421-427. https://doi.org/10.1016/j.conbuildmat.2012.04.013

[46] Cuenca, E., Ferrara, L., Theoretical and Applied Fracture Mechanics, 106, 2020, 102468. https://doi.org/10.1016/j.tafmec.2019.102468

[47] Singh, H., Gupta, R., Case Studies in Construction Materials, 12, 2020, e00324. https://doi.org/10.1016/j.cscm.2019.e00324

[48] Bao, W., Wang, D., Wang, H., Materials China, 38[4] 2019, 396-400.

[49] Asrat, F.S., Ghebrab, T.T., Materials, 13[4] 2020, 840. https://doi.org/10.3390/ma13040840

[50] Wang, X.F., Zhang, J.H., Zhao, W., Han, R., Han, N.X., Xing, F., Construction and Building Materials, 165, 2018, 149-162. https://doi.org/10.1016/j.conbuildmat.2017.12.008

[51] Sahin, O., Yıldırım, G., Sahmaran, M., Indian Concrete Journal, 93[12] 2019, 39-46.

[52] Siad, H., Lachemi, M., Sahmaran, M., Mesbah, H.A., Hossain, K.A., Construction and Building Materials, 176, 2018, 313-322. https://doi.org/10.1016/j.conbuildmat.2018.05.026

[53] In, C.W., Holland, R.B., Kim, J.Y., Kurtis, K.E., Kahn, L.F., Jacobs, L.J., NDT and E International, 57, 2013, 36-44. https://doi.org/10.1016/j.ndteint.2013.03.005

[54] Liu, H., Huang, H., Wu, X., Peng, H., Li, Z., Hu, J., Yu, Q., Cement and Concrete Research, 120, 2019, 198-206. https://doi.org/10.1016/j.cemconres.2019.03.014

[55] Wu, X., Huang, H., Liu, H., Zeng, Z., Li, H., Hu, J., Wei, J., Yu, Q., Composites B, 182, 2020, 107605. https://doi.org/10.1016/j.compositesb.2019.107605

[56] Liu, H., Huang, H., Wu, X., Wang, X., Hu, J., Wei, J., Yu, Q., Construction and Building Materials, 242, 2020, 118148. https://doi.org/10.1016/j.conbuildmat.2020.118148

[57] Cuenca, E., Rigamonti, S., Gastaldo Brac, E., Ferrara, L., Journal of Materials in Civil Engineering, 33[3] 2021, 04020491. https://doi.org/10.1061/(ASCE)MT.1943-5533.0003604

[58] Sugama, T., Pyatina, T., Cement and Concrete Composites, 99, 2019, 1-16. https://doi.org/10.1016/j.cemconcomp.2019.02.011

[59] Yoo, D.Y., Shin, W., Chun, B., Banthia, N., Cement and Concrete Research, 133, 2020, 106091. https://doi.org/10.1016/j.cemconres.2020.106091

[60] Beglarigale, A., Vahedi, H., Eyice, D., Yazlcl, H., Journal of Materials in Civil Engineering, 31[4] 2019, 04019028. https://doi.org/10.1061/(ASCE)MT.1943-5533.0002632

[61] Hooshmand, A., Kianoush, R., Siad, H., Lachemi, M., Moslemi, M., Construction and Building Materials, 267, 2021, 120879. https://doi.org/10.1016/j.conbuildmat.2020.120879

[62] Zha, Y., Yu, J., Wang, R., He, P., Cao, Z., Construction and Building Materials, 190, 2018, 308-316. https://doi.org/10.1016/j.conbuildmat.2018.09.115

[63] Wang, R., Yu, J., Gu, S., He, P., Han, X., Liu, Q., Construction and Building Materials, 236, 2020, 117598. https://doi.org/10.1016/j.conbuildmat.2019.117598

[64] He, P., Yu, J., Wang, R., Du, W., Han, X., Gu, S., Liu, Q., Construction and Building Materials, 246, 2020, 118480. https://doi.org/10.1016/j.conbuildmat.2020.118480

[65] Guo, L., Xu, Y., Chen, B., Chai, L., Ding, C., Fei, C., Journal of the Chinese Ceramic Society, 47[7] 2019, 874-883.

[66] Fang, G., Liu, Y., Qin, S., Ding, W., Zhang, J., Hong, S., Xing, F., Dong, B., Construction and Building Materials, 179, 2018, 336-347. https://doi.org/10.1016/j.conbuildmat.2018.05.193

[67] Suleiman, A.R., Nelson, A.J., Nehdi, M.L., Cement and Concrete Composites, 103, 2019, 49-58. https://doi.org/10.1016/j.cemconcomp.2019.04.026

[68] Jiang, Z., Li, J., Li, W., Cement and Concrete Composites, 103, 2019, 112-120. https://doi.org/10.1016/j.cemconcomp.2019.04.004

[69] Litina, C., Al-Tabbaa, A., Construction and Building Materials, 255, 2020, 119389. https://doi.org/10.1016/j.conbuildmat.2020.119389

[70] Wu, H.L., Du, Y.J., Yu, J., Yang, Y.L., Li, V.C., Science of the Total Environment, 716, 2020, 137095. https://doi.org/10.1016/j.scitotenv.2020.137095

[71] Zhang, P., Dai, Y., Gao, K., Wang, W., Zhao, T., Journal of the Chinese Ceramic Society, 47[11] 2019, 1527-1537.

[72] Zhang, P., Dai, Y., Ding, X., Zhou, C., Xue, X., Zhao, T., Construction and Building Materials, 169, 2018, 705-715. https://doi.org/10.1016/j.conbuildmat.2018.03.032

[73] Choi, H., Inoue, M., Kim, D., Choi, H., Sengoku, R., Materials, 12[15] 2019, 2456. https://doi.org/10.3390/ma12152456

[74] Gilford, J., Hassan, M.M., Rupnow, T., Barbato, M., Okeil, A., Asadi, S., Journal of Materials in Civil Engineering, 26[5] 2014, 886-896. https://doi.org/10.1061/(ASCE)MT.1943-5533.0000892

[75] Jiang, Z., Li, W., Yuan, Z., Yang, Z., Journal of Wuhan University of Technology - Materials Science, 29[5] 2014, 938-944. https://doi.org/10.1007/s11595-014-1024-2

[76] Zhou, S., Zhu, H., Yan, Z., Ju, J.W., Journal of Tongji University, 42[10] 2014, 1467-1472.

[77] Shahid, K.A., Jaafar, M.F.M., Yahaya, F.M., Journal of Mechanical Engineering and Sciences, 7[1] 2014, 1227-1235. https://doi.org/10.15282/jmes.7.2014.22.0120

[78] Byoungsun, P., Young, C.C., Journal of Materials Research and Technology, 8[6] 2019, 6058-6073. https://doi.org/10.1016/j.jmrt.2019.09.080

[79] Park, B., Choi, Y.C., International Journal of Concrete Structures and Materials, 13[1] 2019, 36. https://doi.org/10.1186/s40069-019-0349-9

[80] Rajczakowska, M., Nilsson, L., Habermehl-Cwirzen, K., Hedlund, H., Cwirzen, A., Materials, 12[20] 2019, 3298. https://doi.org/10.3390/ma12203298

[81] Rajczakowska, M., Habermehl-Cwirzen, K., Hedlund, H., Cwirzen, A., Materials, 12[23] 2019, 3926. https://doi.org/10.3390/ma12233926

[82] Qureshi, T., Kanellopoulos, A., Al-Tabbaa, A., Construction and Building Materials, 194, 2019, 266-275. https://doi.org/10.1016/j.conbuildmat.2018.11.027

[83] Qureshi, T., Kanellopoulos, A., Al-Tabbaa, A., Construction and Building Materials, 192, 2018, 768-784. https://doi.org/10.1016/j.conbuildmat.2018.10.143

[84] Dong, B., Wang, Y., Ding, W., Li, S., Han, N., Xing, F., Lu, Y., Construction and Building Materials, 56, 2014, 1-6. https://doi.org/10.1016/j.conbuildmat.2014.01.070

[85] Dong, B., Wang, Y., Fang, G., Han, N., Xing, F., Lu, Y., Cement and Concrete Composites, 56, 2015, 46-50. https://doi.org/10.1016/j.cemconcomp.2014.10.006

[86] Van Tittelboom, K., Tsangouri, E., Van Hemelrijck, D., De Belie, N., Cement and Concrete Composites, 57, 2015, 142-152. https://doi.org/10.1016/j.cemconcomp.2014.12.002

[87] Rahmani, H., Bazrgar, H., Magazine of Concrete Research, 67[9] 2015, 476-486. https://doi.org/10.1680/macr.14.00158

[88] Anglani, G., Tulliani, J.M., Antonaci, P., Materials, 13[5] 2020, 1149. https://doi.org/10.3390/ma13051149

[89] Liu, S., Yang, J., Wang, Z., Rong, H., Zhang, L., Journal of the Chinese Ceramic Society, 43[8] 2015, 1083-1089.

[90] Liu, S., Zhu, D., Guo, S., Materials Review, 30[1] 2016, 108-113.

[91] Perez, G., Erkizia, E., Gaitero, J.J., Kaltzakorta, I., Jiménez, I., Guerrero, A., Materials Chemistry and Physics, 165, 2015, 39-48. https://doi.org/10.1016/j.matchemphys.2015.08.047

[92] Gao, L., Guo, E.D., Liu, Z., Liang, S.J.,) Journal of Beijing University of Technology, 41[8] 2015, 1206-1211.

[93] Gao, L., Guo, E., Zhao, Y., Liu, Z., Liu, S., China Civil Engineering Journal, 49[3] 2016, 98-104.

[94] Ikomaa, H., Kishib, T., Sakaib, Y., Kayondoa, M., Journal of Ceramic Processing Research, 16, 2015, 22-27.

[95] Mostavi, E., Asadi, S., Hassan, M.M., Alansari, M., Journal of Materials in Civil Engineering, 27[12] 2015, 04015035. https://doi.org/10.1061/(ASCE)MT.1943-5533.0001314

[96] Kishi, T., Koide, T., Ahn, T.H., Journal of Ceramic Processing Research, 16, 2015, s63-s73.

[97] Roig-Flores, M., Moscato, S., Serna, P., Ferrara, L., Construction and Building Materials, 86, 2015, 1-11. https://doi.org/10.1016/j.conbuildmat.2015.03.091

[98] Gandhimathi, A., Suji, D., Nature Environment and Pollution Technology, 14[3] 2015, 639-644.

[99] Da Silva, F.B., De Belie, N., Boon, N., Verstraete, W., Construction and Building Materials, 93, 2015, 1034-1041. https://doi.org/10.1016/j.conbuildmat.2015.05.049

[100] Formia, A., Irico, S., Cement International, 13[5] 2015, 70-77.

[101] Guo, Y.C., Qian, J.S., Wang, X., Sun, Y., Materials Research Innovations, 19, 2015, S9212-S9215.

[102] Depaa, R.A.B., Kala, T.F., Indian Journal of Science and Technology, 8[36] 2015, 87644. https://doi.org/10.17485/ijst/2015/v8i36/87644

[103] Milla, J., Hassan, M.M., Rupnow, T., Al-Ansari, M., Arce, G., Transportation Research Record, 2577, 2016, 69-77. https://doi.org/10.3141/2577-09

[104] Yuan, J., Chen, X., He, H.L., Yang, B., Zhu, X.J., Journal of Jilin University - Engineering and Technology, 50[2] 2020, 641-647.

[105] Babu, N.G., Siddiraju, S., International Journal of Civil Engineering and Technology, 7[3] 2016, 398-406.

[106] Van Tittelboom, K., Wang, J., Araújo, M., Snoeck, D., Gruyaert, E., Debbaut, B., Derluyn, H., Cnudde, V., Tsangouri, E., Van Hemelrijck, D., De Belie, N., Construction and Building Materials, 107, 2016, 125-137. https://doi.org/10.1016/j.conbuildmat.2015.12.186

[107] Tan, N.P.B., Keung, L.H., Choi, W.H., Lam, W.C., Leung, H.N., Journal of Applied Polymer Science, 133[12] 2016, 43090. https://doi.org/10.1002/app.43090

[108] Qian, C., Li, R., Luo, M., Chen, H., Journal Wuhan University of Technology - Materials Science, 31[3] 2016, 557-562. https://doi.org/10.1007/s11595-016-1410-z

[109] Roig-Flores, M., Pirritano, F., Serna, P., Ferrara, L., Construction and Building Materials, 114, 2016, 447-457. https://doi.org/10.1016/j.conbuildmat.2016.03.196

[110] Gruyaert, E., Van Tittelboom, K., Sucaet, J., Anrijs, J., Van Vlierberghe, S., Dubruel, P., De Geest, B.G., Remon, J.P., De Belie, N., Materiales de Construccion, 66[323] 2016, e092. https://doi.org/10.3989/mc.2016.07115

[111] Qureshi, T.S., Kanellopoulos, A., Al-Tabbaa, A., Construction and Building Materials, 121, 2016, 629-643. https://doi.org/10.1016/j.conbuildmat.2016.06.030

[112] Alghamri, R., Kanellopoulos, A., Al-Tabbaa, A., Construction and Building Materials, 124, 2016, 910-921. https://doi.org/10.1016/j.conbuildmat.2016.07.143

[113] Souradeep, G., Kua, H.W., Journal of Materials in Civil Engineering, 28[12] 2016, 04016165. https://doi.org/10.1061/(ASCE)MT.1943-5533.0001687

[114] Belleghem, B.V., Van den Heede, P., Tittelboom, K.V., De Belie, N.D., Materials, 10[1] 2017, 5. https://doi.org/10.3390/ma10010005

[115] Zhang, Z., Qian, S., Liu, H., Li, V.C., Transportation Research Record, 2640[1] 2017, 78-83. https://doi.org/10.3141/2640-09

[116] Bonilla, L., Hassan, M., Noorvand, H., Rupnow, T., Okeil, A., Transportation Research Record, 2629[1] 2017, 63-72. https://doi.org/10.3141/2629-09

[117] Dong, B., Fang, G., Wang, Y., Liu, Y., Hong, S., Zhang, J., Lin, S., Xing, F., Cement and Concrete Composites, 78, 2017, 84-96. https://doi.org/10.1016/j.cemconcomp.2016.12.005

[118] Xu, D., Chen, W., Fan, X., Construction and Building Materials, 256, 2020, 119343. https://doi.org/10.1016/j.conbuildmat.2020.119343

[119] Prabahar, A.M., Dhanya, R., Ramasamy, N.G., Dhanasekar, S., Rasayan Journal of Chemistry, 10[2] 2017, 577-583.

[120] Calvo, J.L.G., Pérez, G., Carballosa, P., Erkizia, E., Gaitero, J.J., Guerrero, A., Construction and Building Materials, 138, 2017, 306-315. https://doi.org/10.1016/j.conbuildmat.2017.02.015

[121] Gilabert, F.A., Van Tittelboom, K., Tsangouri, E., Van Hemelrijck, D., De Belie, N., Van Paepegem, W., Cement and Concrete Composites, 79, 2017, 76-93. https://doi.org/10.1016/j.cemconcomp.2017.01.011

[122] Chen, Y., Xia, C., Shepard, Z., Smith, N., Rice, N., Peterson, A.M., Sakulich, A., ACS Sustainable Chemistry and Engineering, 5[5] 2017, 3955-3962. https://doi.org/10.1021/acssuschemeng.6b03142

[123] Liu, Y., Ding, W., Dong, P., Han, S., Li, H., Chen, S., Hong, S., Dong, B., Xing, F., Journal of Functional Materials, 48[7] 2017, 07006-07011.

[124] Dong, B., Ding, W., Qin, S., Fang, G., Liu, Y., Dong, P., Han, S., Xing, F., Hong, S., Construction and Building Materials, 168, 2018, 11-20. https://doi.org/10.1016/j.conbuildmat.2018.02.094

[125] Gardner, D., Herbert, D., Jayaprakash, M., Jefferson, A., Paul, A., Journal of Materials in Civil Engineering, 29[11] 2017, 04017228. https://doi.org/10.1061/(ASCE)MT.1943-5533.0002092

[126] Pérez, G., Calvo, J.L.G., Carballosa, P., Tian, R., Allegro, V.R., Erkizia, E., Gaitero, J.J., Guerrero, A., Magazine of Concrete Research, 69[23] 2017, 1231-1242. https://doi.org/10.1680/jmacr.17.00075

[127] Milla, J., Hassan, M.M., Rupnow, T., Journal of Materials in Civil Engineering, 29[12] 2017, 04017235. https://doi.org/10.1061/(ASCE)MT.1943-5533.0002072

[128] Sumitha, V., Ravichandran, P.T., Krishnan, K.D., International Journal of Engineering and Technology, 7, 2018, 411-414. https://doi.org/10.14419/ijet.v7i2.12.11507

[129] Medjigbodo, S., Bendimerad, A.Z., Rozière, E., Loukili, A., Cement and Concrete Composites, 86, 2018, 72-86. https://doi.org/10.1016/j.cemconcomp.2017.11.003

[130] Bonilla, L., Hassan, M.M., Noorvand, H., Rupnow, T., Okeil, A., Journal of Materials in Civil Engineering, 30[2] 2018, 04017277. https://doi.org/10.1061/(ASCE)MT.1943-5533.0002134

[131] Igawa, H., Eguchi, H., Kitsutaka, Y., Journal of Structural and Construction Engineering, 83[748] 2018, 763-772. https://doi.org/10.3130/aijs.83.763

[132] Escoffres, P., Desmettre, C., Charron, J.P., Construction and Building Materials, 173, 2018, 763-774. https://doi.org/10.1016/j.conbuildmat.2018.04.003

[133] Feng, J., Zhang, P., Chen, W., Yang, J., Liu, H., Journal of Building Materials, 21[4] 2018, 656-662.

[134] Wei, Y., Cheng, P., Liu, M., Zhao, Q., Journal of Building Materials, 21[4] 2018, 588-594.

[135] Mauludin, L.M., Oucif, C., Underground Space, 3[3] 2018, 181-189. https://doi.org/10.1016/j.undsp.2018.04.004

[136] Wang, J., Ding, S., Han, B., Ni, Y.Q., Ou, J., Smart Materials and Structures, 27[11] 2018, 115033. https://doi.org/10.1088/1361-665X/aae59f

[137] Tomczak, K., Jakubowski, J., Construction and Building Materials, 187, 2018, 149-159. https://doi.org/10.1016/j.conbuildmat.2018.07.176

[138] Yuan, L., Chen, S., Wang, S., Huang, Y., Yang, Q., Liu, S., Wang, J., Du, P., Cheng, X., Zhou, Z., Materials, 12[7] 2019, 2818. https://doi.org/10.3390/ma12172818

[139] Du, W., Yu, J., Gu, Y., Li, Y., Han, X., Liu, Q., Construction and Building Materials, 202, 2019, 762-769. https://doi.org/10.1016/j.conbuildmat.2019.01.007

[140] Kang, C., Kim, T., Applied Sciences, 9[8] 2019, 1537. https://doi.org/10.3390/app9081537

[141] Van Mullem, T., Gruyaert, E., Debbaut, B., Caspeele, R., De Belie, N., Construction and Building Materials, 203, 2019, 541-551. https://doi.org/10.1016/j.conbuildmat.2019.01.105

[142] Azarsa, P., Gupta, R., Biparva, A., Cement and Concrete Composites, 99, 2019, 17-31. https://doi.org/10.1016/j.cemconcomp.2019.02.017

[143] Danner, T., Jakobsen, U.H., Geiker, M.R., Minerals, 9[5] 2019, 284.
https://doi.org/10.3390/min9050284

[144] Wang, L., Wu, S., Journal of Hydraulic Engineering, 50[7] 2019, 787-797.
https://doi.org/10.1109/TCYB.2018.2873733

[145] Wang, X., Huang, Y., Huang, Y., Zhang, J., Fang, C., Yu, K., Chen, Q., Li, T., Han,
R., Yang, Z., Xu, P., Liang, G., Su, D., Ding, X., Li, D., Han, N., Xing, F.,
Construction and Building Materials, 220, 2019, 90-101.
https://doi.org/10.1016/j.conbuildmat.2019.06.017

[146] Van Belleghem, B., Zaccardi, Y.V., Van den Heede, P., Van Tittelboom, K., De Belie,
N., Construction and Building Materials, 227, 2019, 116789.
https://doi.org/10.1016/j.conbuildmat.2019.116789

[147] Abro, F.R., Buller, A.S., Lee, K.M., Jang, S.Y., Materials, 12[11] 2019, 1865.
https://doi.org/10.3390/ma12111865

[148] Lv, L., Guo, P., Liu, G., Han, N., Xing, F., Cement and Concrete Composites, 105,
2020, 103445. https://doi.org/10.1016/j.cemconcomp.2019.103445

[149] Mullem, T., Gruyaert, E., Caspeele, R., Belie, N., Materials, 13[4] 2020, 0997.

[150] De Nardi, C., Gardner, D., Jefferson, A.D., Materials, 13[6] 2020, 1328.
https://doi.org/10.3390/ma13061328

[151] Li, Z., Souza, L.R.D., Litina, C., Markaki, A.E., Al-Tabbaa, A., Materials and Design,
190, 2020, 108572. https://doi.org/10.1016/j.matdes.2020.108572

[152] Selvarajoo, T., Davies, R.E., Freeman, B.L., Jefferson, A.D., Construction and
Building Materials, 254, 2020, 119245.
https://doi.org/10.1016/j.conbuildmat.2020.119245

[153] Kumar, C.N., Prabhakar, M.N., Song, J.I., Polymer Composites, 41[5] 2020, 1913-
1924. https://doi.org/10.1002/pc.25507

[154] Selvarajoo, T., Davies, R.E., Gardner, D.R., Freeman, B.L., Jefferson, A.D.,
Construction and Building Materials, 245, 2020, 118332.
https://doi.org/10.1016/j.conbuildmat.2020.118332

[155] Lefever, G., Snoeck, D., Aggelis, D.G., De Belie, N., Van Vlierberghe, S., Van
Hemelrijck, D., Materials, 13[2] 2020, 380. https://doi.org/10.3390/ma13020380

[156] Hu, M., Guo, J., Du, J., Liu, Z., Li, P., Ren, X., Feng, Y., Journal of Colloid and
Interface Science, 538, 2019, 397-403. https://doi.org/10.1016/j.jcis.2018.12.004

[157] Snoeck, D., De Schryver, T., De Belie, N., Construction and Building Materials, 191,
2018, 13-22. https://doi.org/10.1016/j.conbuildmat.2018.10.015

[158] Kanellopoulou, I., Karaxi, E.K., Karatza, A., Kartsonakis, I.A., Charitidis, C.A., Fatigue and Fracture of Engineering Materials and Structures, 42[7] 2019, 1494-1509. https://doi.org/10.1111/ffe.12998

[159] Sidiq, A., Gravina, R., Setunge, S., Giustozzi, F., Construction and Building Materials, 253, 2020, 119175. https://doi.org/10.1016/j.conbuildmat.2020.119175

[160] Xue, C., Li, W., Li, J., Tam, V.W.Y., Ye, G., Structural Concrete, 20[1] 2019, 198-212. https://doi.org/10.1002/suco.201800177

[161] Hilloulin, B., Van Tittelboom, K., Gruyaert, E., De Belie, N., Loukili, A., Cement and Concrete Composites, 55, 2015, 298-307. https://doi.org/10.1016/j.cemconcomp.2014.09.022

[162] Kanellopoulos, A., Giannaros, P., Palmer, D., Kerr, A., Al-Tabbaa, A., Smart Materials and Structures, 26[4] 2017, 045025. https://doi.org/10.1088/1361-665X/aa516c

[163] Feiteira, J., Gruyaert, E., De Belie, N., Construction and Building Materials, 102, 2016, 671-678. https://doi.org/10.1016/j.conbuildmat.2015.10.192

[164] Šavija, B., Feiteira, J., Araújo, M., Chatrabhuti, S., Raquez, J.M., Van Tittelboom, K., Gruyaert, E., De Belie, N., Schlangen, E., Materials, 10[1] 2017, 10. https://doi.org/10.3390/ma10010010

[165] Gilabert, F.A., Van Tittelboom, K., Van Stappen, J., Cnudde, V., De Belie, N., Van Paepegem, W., Cement and Concrete Composites, 77, 2017, 68-80. https://doi.org/10.1016/j.cemconcomp.2016.12.001

[166] Van Belleghem, B., Kessler, S., Van den Heede, P., Van Tittelboom, K., De Belie, N., Cement and Concrete Research, 113, 2018, 130-139. https://doi.org/10.1016/j.cemconres.2018.07.009

[167] Hu, Z.X., Hu, X.M., Cheng, W.M., Zhao, Y.Y., Wu, M.Y., Construction and Building Materials, 179, 2018, 151-159. https://doi.org/10.1016/j.conbuildmat.2018.05.199

[168] Wu, S., Lu, G., Liu, Q., Liu, P., Yang, J., Advances in Materials Science and Engineering, 2020, 2020, 5395602.

[169] Chindasiriphan, P., Yokota, H., Pimpakan, P., Construction and Building Materials, 233, 2020, 116975. https://doi.org/10.1016/j.conbuildmat.2019.116975

[170] Taheri, S., Clark, S.M., International Journal of Concrete Structures and Materials, 15[1] 2021, 8. https://doi.org/10.1186/s40069-020-00449-2

[171] Taheri, M.N., Sabet, S.A., Kolahchi, R., Smart Structures and Systems, 25[3] 2020, 337-343.

[172] Feng, J., Dong, H., Wang, R., Su, Y., Cement and Concrete Research, 133, 2020, 106053. https://doi.org/10.1016/j.cemconres.2020.106053

[173] Alghamri, R., Al-Tabbaa, A., Construction and Building Materials, 254, 2020, 119254. https://doi.org/10.1016/j.conbuildmat.2020.119254

[174] Han, T., Wang, X., Li, D., Li, D., Xing, F., Ren, J., Han, N., Construction and Building Materials, 241, 2020, 118009. https://doi.org/10.1016/j.conbuildmat.2020.118009

[175] Du, W., Yu, J., Gu, S., Wang, R., Li, J., Han, X., Liu, Q., Construction and Building Materials, 247, 2020, 118575. https://doi.org/10.1016/j.conbuildmat.2020.118575

[176] Ren, Y., Abbas, N., Zhu, G., Tang, J., Colloids and Surfaces A, 587, 2020, 124347. https://doi.org/10.1016/j.colsurfa.2019.124347

[177] Wang, X., Chen, Z., Xu, W., Wang, X., Composites B, 184, 2020, 107744. https://doi.org/10.1016/j.compositesb.2020.107744

[178] Zhuang, X., Nguyen-Xuan, H., Zhou, S., Computers, Materials and Continua, 67[1] 2021, 577-593. https://doi.org/10.32604/cmc.2021.014688

[179] Wang, M., Hu, X., Zhao, Y., Advances in Structural Engineering, 24[1] 2021, 52-64. https://doi.org/10.1177/1369433220942868

[180] Xue, C., Li, W., Wang, K., Sheng, D., Shah, S.P., Shah, S.P., Smart Materials and Structures, 29[8] 2020, 085004. https://doi.org/10.1088/1361-665X/ab8eb6

[181] Mauludin, L.M., Oucif, C., Rabczuk, T., Frontiers of Structural and Civil Engineering, 14[3] 2020, 792-801. https://doi.org/10.1007/s11709-020-0629-0

[182] Peng, Z., Yu, C., Feng, Q., Zheng, Y., Huo, J., Liu, X., Energy Sources A, 2019,

[183] Nie, F.M., Cui, J., Zhou, Y.F., Pan, L., Ma, Z., Li, Y.S., Macromolecules, 52[14] 2019, 5289-5297. https://doi.org/10.1021/acs.macromol.9b00871

[184] Jonkers, H.M., Thijssen, A., Muyzer, G., Copuroglu, O., Schlangen, E., Ecological Engineering, 36[2] 2010, 230-235. https://doi.org/10.1016/j.ecoleng.2008.12.036

[185] Wiktor, V., Jonkers, H.M., Materiaux et Techniques, 99[5] 2011, 565-571. https://doi.org/10.1051/mattech/2011110

[186] Wiktor, V., Jonkers, H.M., Cement and Concrete Composites, 33[7] 2011, 763-770. https://doi.org/10.1016/j.cemconcomp.2011.03.012

[187] Wang, J., Van Tittelboom, K., De Belie, N., Verstraete, W., Construction and Building Materials, 26[1] 2012, 532-540. https://doi.org/10.1016/j.conbuildmat.2011.06.054

[188] Wang, J.Y., De Belie, N., Verstraete, W., Journal of Industrial Microbiology and Biotechnology, 39[4] 2012, 567-577. https://doi.org/10.1007/s10295-011-1037-1

[189] Qian, C., Li, R., Pan, Q., Luo, M., Rong, H., Journal of Southeast University - Natural Science, 43[2] 2013, 360-364.

[190] Qian, C., Luo, M., Pan, Q., Li, R., Journal of the Chinese Ceramic Society, 41[5] 2013, 620-626.

[191] Gao, L., Sun, G., Journal of the Chinese Ceramic Society, 41[5] 2013, 627-636.

[192] Zemskov, S.V., Jonkers, H.M., Vermolen, F.J., Journal of Intelligent Material Systems and Structures, 25[1] 2014, 4-12. https://doi.org/10.1177/1045389X12437887

[193] Talaiekhozani, A., Keyvanfar, A., Andalib, R., Samadi, M., Shafaghat, A., Kamyab, H., Majid, M.Z.A., Zin, R.M., Fulazzaky, M.A., Lee, C.T., Hussin, M.W., Desalination and Water Treatment, 52[19-21] 2014, 3623-3630. https://doi.org/10.1080/19443994.2013.854092

[194] Ke, J., Peng, H., Liu, B., Deng, X., Xing, F., Journal of Shenzhen University - Science and Engineering, 32[2] 2015, 145-151. https://doi.org/10.3724/SP.J.1249.2015.02145

[195] Zhang, J., Liu, Y., Feng, T., Zhou, M., Zhao, L., Zhou, A., Li, Z., Construction and Building Materials, 148, 2017, 610-617. https://doi.org/10.1016/j.conbuildmat.2017.05.021

[196] Alazhari, M., Sharma, T., Heath, A., Cooper, R., Paine, K., Construction and Building Materials, 160, 2018, 610-619. https://doi.org/10.1016/j.conbuildmat.2017.11.086

[197] Jiang, L., Jia, G., Wang, Y., Li, Z., ACS Applied Materials and Interfaces, 12[9] 2020, 10938-10948. https://doi.org/10.1021/acsami.9b21465

[198] Ganesh, G.M., Santhi, A.S., Kalaichelvan, G., International Journal of Civil Engineering and Technology, 8[9] 2017, 539-545.

[199] Zhang, J., Xu, S., Feng, T., Zhao, L., Li, Z., Journal of Tsinghua University, 59[8] 2019, 607-613.

[200] Jiang, L., Jia, G., Jiang, C., Li, Z., Construction and Building Materials, 232, 2020, 117222. https://doi.org/10.1016/j.conbuildmat.2019.117222

[201] Xu, J., Wang, X., Construction and Building Materials, 167, 2018, 1-14. https://doi.org/10.1016/j.conbuildmat.2018.02.020

[202] Xu, J., Wang, X., Journal of Tsinghua University, 59[8] 2019, 601-606.

[203] Zheng, T., Su, Y., Zhang, X., Zhou, H., Qian, C., ACS Applied Materials and Interfaces, 12[47] 2020, 52415-52432. https://doi.org/10.1021/acsami.0c16343

[204] Su, Y., Zheng, T., Qian, C., Construction and Building Materials, 273, 2021, 121740. https://doi.org/10.1016/j.conbuildmat.2020.121740

[205] Xu, H., Lian, J., Gao, M., Fu, D., Yan, Y., Materials, 12[14] 2019, 2313. https://doi.org/10.3390/ma12142313

[206] Gupta, S., Kua, H.W., Pang, S.D., Cement and Concrete Composites, 86, 2018, 238-254. https://doi.org/10.1016/j.cemconcomp.2017.11.015

[207] Kua, H.W., Gupta, S., Aday, A.N., Srubar, W.V., Cement and Concrete Composites, 100, 2019, 35-52. https://doi.org/10.1016/j.cemconcomp.2019.03.017

[208] Wang, J., Mignon, A., Snoeck, D., Wiktor, V., Van Vliergerghe, S., Boon, N., De Belie, N., Frontiers in Microbiology, 6[10] 2015, 1088. https://doi.org/10.3389/fmicb.2015.01088

[209] Pungrasmi, W., Intarasoontron, J., Jongvivatsakul, P., Likitlersuang, S., Scientific Reports, 9[1] 2019, 12484. https://doi.org/10.1038/s41598-019-49002-6

[210] Khaliq, W., Ehsan, M.B., Construction and Building Materials, 102, 2016, 349-357. https://doi.org/10.1016/j.conbuildmat.2015.11.006

[211] Sarkar, M., Adak, D., Tamang, A., Chattopadhyay, B., Mandal, S., RSC Advances, 5[127] 2015, 105363-105371. https://doi.org/10.1039/C5RA20858K

[212] Nguyen, T.H., Ghorbel, E., Fares, H., Cousture, A., Cement and Concrete Composites, 104, 2019, 103340. https://doi.org/10.1016/j.cemconcomp.2019.103340

[213] Ling, H., Qian, C., Construction and Building Materials, 144, 2017, 406-411. https://doi.org/10.1016/j.conbuildmat.2017.02.160

[214] He, H., Li, G., Zhang, J., Zhang, J., Luo, M., Hu, W., Lin, Y., Deng, Z., Liu, Z., Chen, W., Deng, X., Journal of Ceramic Processing Research, 20[5] 2019, 470-478. https://doi.org/10.36410/jcpr.2019.20.5.470

[215] Ganesh, A.C., Muthukannan, M., Malathy, R., Babu, C.R., KSCE Journal of Civil Engineering, 23[10] 2019, 4368-4377. https://doi.org/10.1007/s12205-019-1661-2

[216] Shaheen, N., Khushnood, R.A., Khaliq, W., Murtaza, H., Iqbal, R., Khan, M.II., Construction and Building Materials, 226, 2019, 492-506. https://doi.org/10.1016/j.conbuildmat.2019.07.202

[217] Nkuah, J.S., Kaushal, M., Singla, S., International Journal of Scientific and Technology Research, 9[2] 2020, 1553-1567.

[218] Khushnood, R.A., Shaheen, N., Ahmad, S., Zarrar, F., Journal of Intelligent Material Systems and Structures, 30[1] 2019, 3-15. https://doi.org/10.1177/1045389X18806401

[219] Seifan, M., Sarmah, A.K., Samani, A.K., Ebrahiminezhad, A., Ghasemi, Y., Berenjian, A., Applied Microbiology and Biotechnology, 102[10] 2018, 4489-4498. https://doi.org/10.1007/s00253-018-8913-9

[220] Seifan, M., Sarmah, A.K., Ebrahiminezhad, A., Ghasemi, Y., Samani, A.K., Berenjian, A., Applied Microbiology and Biotechnology, 102[5] 2018, 2167-2178. https://doi.org/10.1007/s00253-018-8782-2

[221] Seifan, M., Ebrahiminezhad, A., Ghasemi, Y., Samani, A.K., Berenjian, A., Applied Microbiology and Biotechnology, 102[1] 2018, 175-184. https://doi.org/10.1007/s00253-017-8611-z

[222] Khushnood, R.A., Qureshi, Z.A., Shaheen, N., Ali, S., Science of the Total Environment, 703, 2020, 135007. https://doi.org/10.1016/j.scitotenv.2019.135007

[223] Singh, H., Gupta, R., Journal of Building Engineering, 28, 2020, 101090. https://doi.org/10.1016/j.jobe.2019.101090

[224] Hamza, O., Esaker, M., Elliott, D., Souid, A., Materials Today Communications, 24, 2020, 100988. https://doi.org/10.1016/j.mtcomm.2020.100988

[225] Rauf, M., Khaliq, W., Khushnood, R.A., Ahmed, I., Construction and Building Materials, 258, 2020, 119578. https://doi.org/10.1016/j.conbuildmat.2020.119578

[226] Su, Y., Qian, C., Rui, Y., Feng, J., Cement and Concrete Composites, 116, 2021, 103896. https://doi.org/10.1016/j.cemconcomp.2020.103896

[227] Vijay, K., Murmu, M., Frontiers of Structural and Civil Engineering, 13[3] 2019, 515-525. https://doi.org/10.1007/s11709-018-0494-2

[228] Huynh, N.N.T., Imamoto, K.I., Kiyohara, C., Journal of Advanced Concrete Technology, 17[12] 2019, 700-714. https://doi.org/10.3151/jact.17.700

[229] Sun, X., Miao, L., Wu, L., Wang, C., Chen, R., Advances in Cement Research, 32[6] 2020, 262-272. https://doi.org/10.1680/jadcr.18.00121

[230] Shahid, S., Aslam, M.A., Ali, S., Zameer, M., Faisal, M., ChemistrySelect, 5[1] 2020, 312-323. https://doi.org/10.1002/slct.201902206

[231] Prayuda, H., Soebandono, B., Cahyati, M.D., Monika, F., International Journal of Integrated Engineering, 12[4] 2020, 300-309.

[232] Mondal, S., Ghosh, A.D., Construction and Building Materials, 266, 2021, 121122. https://doi.org/10.1016/j.conbuildmat.2020.121122

[233] Mondal, S., Das, P., Datta, P., Ghosh, A.D., Cement and Concrete Composites, 108, 2020, 103523. https://doi.org/10.1016/j.cemconcomp.2020.103523

[234] Pérez, H.F., García, M.G., Journal of Applied Research and Technology, 18[5] 2020, 245-258.

[235] Wang, J., Jonkers, H.M., Boon, N., De Belie, N., Applied Microbiology and Biotechnology, 101[12] 2017, 5101-5114. https://doi.org/10.1007/s00253-017-8260-2

[236] Williams, S.L., Kirisits, M.J., Ferron, R.D., Construction and Building Materials, 139, 2017, 611-618. https://doi.org/10.1016/j.conbuildmat.2016.09.155

[237] Metwally, G.A.M., Mahdy, M., El-Raheem, A.H.A., Civil Engineering Journal (Iran), 6[8] 2020, 1443-1456. https://doi.org/10.28991/cej-2020-03091559

[238] Palin, D., Wiktor, V., Jonkers, H.M., Smart Materials and Structures, 25[8] 2016, 084008. https://doi.org/10.1088/0964-1726/25/8/084008

[239] Zhang, J.L., Wu, R.S., Li, Y.M., Zhong, J.Y., Deng, X., Liu, B., Han, N.X., Xing, F., Applied Microbiology and Biotechnology, 100[15] 2016, 6661-6670. https://doi.org/10.1007/s00253-016-7382-2

[240] Zhang, J.L., Wang, C.G., Wang, Q.L., Feng, J.L., Pan, W., Zheng, X.C., Liu, B., Han, N.X., Xing, F., Deng, X., Applied Microbiology and Biotechnology, 100[24] 2016, 10295-10306. https://doi.org/10.1007/s00253-016-7741-z

[241] Zhang, J., Mai, B., Cai, T., Luo, J., Wu, W., Liu, B., Han, N., Xing, F., Deng, X., Materials, 10[2] 2017, 116. https://doi.org/10.3390/ma10020116

[242] Tziviloglou, E., Wiktor, V., Jonkers, H.M., Schlangen, E., Construction and Building Materials, 122, 2016, 118-125. https://doi.org/10.1016/j.conbuildmat.2016.06.080

[243] Wang, J., Dewanckele, J., Cnudde, V., Van Vlierberghe, S., Verstraete, W., De Belie, N., Cement and Concrete Composites, 53, 2014, 289-304. https://doi.org/10.1016/j.cemconcomp.2014.07.014

[244] Wang, J.Y., Snoeck, D., Van Vlierberghe, S., Verstraete, W., De Belie, N., Construction and Building Materials, 68, 2014, 110-119. https://doi.org/10.1016/j.conbuildmat.2014.06.018

[245] Gao, M., Guo, J., Cao, H., Wang, H., Xiong, X., Krastev, R., Nie, K., Xu, H., Liu, L., Journal of Environmental Management, 261, 2020, 110225. https://doi.org/10.1016/j.jenvman.2020.110225

[246] Xu, J., Yao, W., Cement and Concrete Research, 64, 2014, 1-10. https://doi.org/10.1016/j.cemconres.2014.06.003

[247] Luo, M., Qian, C.X., Journal of Materials in Civil Engineering, 28[12] 2016, 04016151. https://doi.org/10.1061/(ASCE)MT.1943-5533.0001673

[248] Wang, J.Y., Soens, H., Verstraete, W., De Belie, N., Cement and Concrete Research, 56, 2014, 139-152. https://doi.org/10.1016/j.cemconres.2013.11.009

[249] Xu, J., Wang, X., Wang, B., Applied Microbiology and Biotechnology, 102[7] 2018, 3121-3132. https://doi.org/10.1007/s00253-018-8779-x

[250] Xu, J., Wang, B., Materials Review, 31[7] 2017, 127-131.

[251] Wang, X., Fang, C., Li, D., Han, N., Xing, F., Cement and Concrete Composites, 92, 2018, 216-229. https://doi.org/10.1016/j.cemconcomp.2018.05.013

[252] Wang, X., Li, W., Jiang, Z., Materials, 13[3] 2020, 644. https://doi.org/10.3390/ma13030644

[253] Wang, X.F., Yang, Z.H., Fang, C., Wang, W., Liu, J., Xing, F., Construction and Building Materials, 235, 2020, 117442. https://doi.org/10.1016/j.conbuildmat.2019.117442

[254] Xu, J., Wang, X., Zuo, J., Liu, X., Advances in Materials Science and Engineering, 2018, 2018, 5153041. https://doi.org/10.1155/2018/5153041

[255] Su, Y., Li, F., He, Z., Qian, C., Journal of Building Engineering, 35, 2021, 102082. https://doi.org/10.1016/j.jobe.2020.102082

[256] Xu, J., Tang, Y., Wang, X., Wang, Z., Yao, W., Construction and Building Materials, 265, 2020, 120364. https://doi.org/10.1016/j.conbuildmat.2020.120364

[257] Ersan, Y.C., Palin, D., Tasdemir, S.B.Y., Tasdemir, K., Jonkers, H.M., Boon, N., De Belie, N., Frontiers in Built Environment, 4, 2018, 70. https://doi.org/10.3389/fbuil.2018.00070

[258] Algaifi, H.A., Bakar, S.A., Sam, A.R.M., Abidin, A.R.Z., Shahir, S., Al-Towayti, W.A.H., Construction and Building Materials, 189, 2018, 816-824. https://doi.org/10.1016/j.conbuildmat.2018.08.218

[259] Algaifi, H.A., Bakar, S.A., Sam, A.R.M., Ismail, M., Abidin, A.R.Z., Shahir, S., Altowayti, W.A.H., Construction and Building Materials, 254, 2020, 119258. https://doi.org/10.1016/j.conbuildmat.2020.119258

[260] David, M.F., Kumar, G.M., Raambalaji, C.B., Subaranjani, S., Muthusivashankar, A., Bose, J.C., Asian Journal of Civil Engineering, 22[1] 2021, 59-71. https://doi.org/10.1007/s42107-020-00298-0

[261] Vermeer, C.M., Rossi, E., Tamis, J., Jonkers, H.M., Kleerebezem, R., Resources, Conservation and Recycling, 164, 2021, 105206. https://doi.org/10.1016/j.resconrec.2020.105206

[262] Alshalif, A.F., Irwan, J.M., Othman, N., Al-Gheethi, A.A., Shamsudin, S., Nasser, I.M., Journal of CO_2 Utilization, 44, 2021, 101412. https://doi.org/10.1016/j.jcou.2020.101412

[263] He, J., Gray, K., Norris, A., Ewing, A.C., Jurgerson, J., Shi, X., Journal of Transportation Engineering B, 146[3] 2020, 04020036. https://doi.org/10.1061/JPEODX.0000188

[264] Irwan, J.M., Anneza, L.H., Othman, N., International Journal of Sustainable Construction Engineering and Technology, 10[1] 2019, 80-92. https://doi.org/10.30880/ijscet.2019.10.01.008

[265] Jeong, B., Jho, E.H., Kim, S.H., Nam, K., Journal of Materials in Civil Engineering, 31[10] 2019, 04019227. https://doi.org/10.1061/(ASCE)MT.1943-5533.0002711

[266] Luo, S., Liu, B., Zhang, J., Li, Q., Deng, X., Journal of Shenzhen University - Science and Engineering, 37[1] 2020, 25-32.

[267] Jung, Y., Kim, W., Kim, W., Park, W., Journal of Microbiology and Biotechnology, 30[3] 2020, 404-416. https://doi.org/10.4014/jmb.1908.08044

[268] Ryu, Y., Lee, K.E., Cha, I.T., Park, W., Journal of Microbiology, 58[4] 2020, 288-296. https://doi.org/10.1007/s12275-020-9580-y

[269] Han, S., Choi, E.K., Park, W., Yi, C., Chung, N., Applied Biological Chemistry, 62[1] 2019, 19. https://doi.org/10.1186/s13765-019-0426-4

[270] Kalenov, S.V., Gradova, N.B., Sivkov, S.P., Agalakova, E.V., Belov, A.A., Suyasov, N.A., Khokhlachev, N.S., Panfilov, V.I., Biotekhnologiya, 36[4] 2020, 21-28. https://doi.org/10.21519/0234-2758-2020-36-4-21-28

[271] Son, H.M., Kim, H.Y., Park, S.M., Lee, H.K., Materials, 11[5] 2018, 782. https://doi.org/10.3390/ma11050782

[272] Rong, H., Wei, G., Ma, G., Zhang, Y., Zheng, X., Zhang, L., Xu, R., Construction and Building Materials, 244, 2020, 118372. https://doi.org/10.1016/j.conbuildmat.2020.118372

[273] Tayebani, B., Mostofinejad, D., Construction and Building Materials, 208, 2019, 75-86. https://doi.org/10.1016/j.conbuildmat.2019.02.172

[274] Vijay, K., Murmu, M., International Journal of Structural Engineering, 10[3] 2020, 217-231. https://doi.org/10.1504/IJSTRUCTE.2020.10029530

[275] Lu, S., Chen, M., Dang, Y., Cao, L., He, J., Zhong, J., AIP Advances, 9[10] 2019, 105312. https://doi.org/10.1063/1.5124315

[276] Zamani, M., Nikafshar, S., Mousa, A., Behnia, A., Construction and Building Materials, 249, 2020, 118556. https://doi.org/10.1016/j.conbuildmat.2020.118556

[277] Nielsen, S.D., Paegle, I., Borisov, S.M., Kjeldsen, K.U., Røy, H., Skibsted, J., Koren, K., ACS Omega, 4[23] 2019, 20237-20243. https://doi.org/10.1021/acsomega.9b02541

[278] Brasileiro, P.P.F., Da Silva, R.C.F.S., Rocha E Silva, F.C.P., Brandão, Y.D., Sarubbo, L.A., Benachour, M., Chemical Engineering Transactions, 79, 2020, 97-102.

[279] Müller, W.E.G., Tolba, E., Wang, S., Li, Q., Neufurth, M., Ackermann, M., Muñoz-Espí, R., Schröder, H.C., Wang, X., International Journal of Molecular Sciences, 20[12] 2019, 2948. https://doi.org/10.3390/ijms20122948

[280] Kadapure, S.A., Kulkarni, G.S., Prakash, K.B., International Journal of Sustainable Building Technology and Urban Development, 8[1] 2017, 54-65.

[281] Yunus, B.D., Schlangen, E., Jonkers, H.M., International Journal of Integrated Engineering, 11[9S] 2019, 247-254.

[282] Chen, X., Yuan, J., Alazhari, M., Materials, 12[8] 2019, 1303. https://doi.org/10.3390/ma12081303

[283] Xu, J., Tang, Y., Wang, X., Process Biochemistry, 94, 2020, 266-272. https://doi.org/10.1016/j.procbio.2020.04.028

[284] Khan, M.B.E., Shen, L., Dias-da-Costa, D., Construction and Building Materials, 277, 2021, 122332. https://doi.org/10.1016/j.conbuildmat.2021.122332

[285] Zhang, Y., Qian, C., Zhang, X., Journal of Southeast University - Natural Science, 50[1] 2020, 101-108.

[286] Jin, C., Yu, R., Shui, Z., Frontiers in Built Environment, 4, 2018, 62. https://doi.org/10.3389/fbuil.2018.00062

[287] Luo, J., Chen, X., Crump, J., Zhou, H., Davies, D.G., Zhou, G., Zhang, N., Jin, C., Construction and Building Materials, 164, 2018, 275-285. https://doi.org/10.1016/j.conbuildmat.2017.12.233

[288] Menon, R.R., Luo, J., Chen, X., Zhou, H., Liu, Z., Zhou, G., Zhang, N., Jin, C., Scientific Reports, 9[1] 2019, 2075. https://doi.org/10.1038/s41598-019-39156-8

[289] Zhang, X., Fan, X., Li, M., Samia, A., Yu, X.B., Journal of Cleaner Production, 292, 2021, 125870. https://doi.org/10.1016/j.jclepro.2021.125870

www.ingramcontent.com/pod-product-compliance
Lightning Source LLC
Chambersburg PA
CBHW071704210326
41597CB00017B/2328